METALS
REFERENCE BOOK

METALS

REFERENCE BOOK

VOLUME II

COLIN J. SMITHELLS
M.C., D.Sc., F.I.M.

FOURTH EDITION

LONDON
BUTTERWORTHS
1967

ENGLAND: BUTTERWORTH & CO. (PUBLISHERS) LTD.
                LONDON: 88 Kingsway, W.C.2

AUSTRALIA: BUTTERWORTH & CO. (AUSTRALIA) LTD.
                SYDNEY: 20 Loftus Street
                MELBOURNE: 473 Bourke Street
                BRISBANE: 240 Queen Street

CANADA: BUTTERWORTH & CO. (CANADA) LTD.
                TORONTO: 1367 Danforth Avenue, 6

NEW ZEALAND: BUTTERWORTH & CO. (NEW ZEALAND) LTD.
                WELLINGTON: 49/51 Ballance Street
                AUCKLAND: 35 High Street

SOUTH AFRICA: BUTTERWORTH & CO. (SOUTH AFRICA) LTD.
                DURBAN: 33/35 Beach Grove

U.S.A.:   BUTTERWORTH INC.
                WASHINGTON, D.C.: 7300 Pearl Street, 20014

First Edition 1949
Second Edition 1955
Third Edition 1962
Fourth Edition 1967

©

Butterworth & Co. (Publishers) Limited
1967

Suggested U.D.C. no: 669(083)
Suggested additional nos: 669.017(083)
621.7(083)

Set in Monotype Old Style Type
Printed in Great Britain by Richard Clay (*The Chaucer Press*), Ltd., Bungay, Suffolk

PREFACE TO THE FOURTH EDITION

THE OBJECT of this *Reference Book* is to provide a convenient summary of data relating to metallurgy and metal physics. So far as possible the data are presented in the form of tables or diagrams with the minimum of descriptive matter, although short monographs have been included where information could not otherwise be adequately presented.

The values given are those which contributors have selected as the most reliable after a critical review of the published data. It is believed that the contributors' choice is likely to be more reliable than that which the reader might make from a series of conflicting values, especially if he is not a specialist in the field in question. A bibliography at the end of each chapter enables the reader to refer to the more important original sources.

The data throughout the book have been completely revised and many of the original sections have been rewritten. A new section on the application of Lasers in metallurgy, and a table of properties of Elementary Particles have been introduced.

The Editor takes this opportunity of thanking those readers who have drawn attention to errors, or who have made helpful suggestions in regard to later editions. He would again express his gratitude to the contributors whose names appear below for their help and co-operation, and to many others who have given generous assistance in the preparation of this edition. He would also like to express his appreciation of the patience and co-operation shown by both the publishers and printers.

C. J. S.

Chalfont St. Peter, Bucks

CONTENTS

VOLUME I

VOLUME II

VOLUME III

Editor

Colin J. Smithells, M.C., D.Sc., F.I.M.

Contributors to this edition

C. R. Barber, B.Sc., F.Inst.P.

E. R. Buckle, Ph.D., F.R.I.C.

J. R. Butler, M.A., Ph.D.

F. A. Champion, A.R.C.S., Ph.D., F.I.M.

A. R. L. Chivers, M.A.

A. D. LeClaire, B.A.

H. R. Clayton, M.Sc., F.R.I.C.

B. L. Daniell, B.Sc., Ph.D., D.C.T.

A. E. Dodd, M.B.E., Ph.D., M.Sc., F.R.I.C.

D. M. Dovey, M.A., Ph.D., A.R.I.C.

R. Eborall, M.A.

D. Edge, B.Sc.

A. H. Edwards, B.Sc., A.R.I.C., M.Inst.F.

E. F. Emley, Ph.D., F.R.I.C., F.I.M.

K. M. Entwistle, Ph.D., M.Sc., F.I.M.

W. P. Fentiman, B.Sc.

I. Fitzpatrick, M.Sc., Ph.D.

M. L. H. Flindt, M.B., B.S., L.R.C.P., M.R.C.S.

P. G. Forrester, M.Sc.

G. A. Geach, M.Sc., Ph.D., F.I.M.

D. N. Gwyther, B.Sc., C.Eng., M.I.GasE., M.Inst.F.

M. A. Haughton, M.A.Sc., A.R.S.M., A.I.M.

W. F. Higgins, M.Sc., Ph.D., A.R.I.C.

J. Hinde, F.I.M., M.Inst.W., A.R.Ae.S.

A. G. L. E. Hock, B.Sc., Ph.D.(Shef.), F.R.I.C.

N. P. Inglis, Ph.D., M.Eng., M.I.Mech.E., F.I.M.

W. J. Jackson, B.Sc., M.Sc.(Eng.), A.R.I.C., F.I.M.

I. Jenkins, D.Sc., M.Sc., F.I.M.

R. E. Kemp, B.Sc.

O. Kubaschewski, D.Phil. Habil.

D. S. Laidler, B.Sc., Ph.D., A.R.I.C.

L. A. J. Lodder, B.Sc., A.R.C.S.

E. Marks, M.Sc., Ph.D.

J. J. McGlynn, M.A.

A. B. McIntosh, Ph.D., F.R.I.C., A.R.C.S.T., F.I.M.

A. D. Morgan

W. R. E. Nice, B.Sc., A.R.S.M., A.I.M.

L. B. Pfeil, O.B.E., D.Sc., A.R.S.M., F.I.M., F.R.S.

H. W. L. Phillips, M.A., F.R.I.C., F.Inst.P., F.I.M.

H. P. Rooksby, B.Sc., F.Inst.P.

B. A. Scott, Ph.D., A.R.C.S., F.R.I.C., F.I.M.

C. Gordon Smith, M.A., A.M.I.E.E.

D. Tabor, Ph.D.(Cantab.).

D. E. J. Talbot, M.Sc.

[*contd*]

Contributors to this edition (continued)

P. E. B. Vaile, A.M.I.Mech.E., F.Inst.Pet., M.S.A.E.

P. C. Varley, M.B.E., T.D., M.A., F.I.M.

T. G. Walker, B.Sc., Ph.D.

M. Whyte, B.Sc.

G. W. Wilson, B.Sc., Ph.D.

D. A. Wright, D.Sc., M.Sc., F.Inst.P.

J. G. Young, B.Sc.

ACKNOWLEDGEMENTS

THE Editor and the Publishers desire to thank all those who have kindly authorized the reproduction of diagrams and tables and in particular the following :

American Ceramic Society, Columbus, Ohio
American Institute of Mining and Metallurgical Engineers, New York
American Society for Metals, Cleveland, Ohio
American Society for Testing Materials, Philadelphia, Pa
Cavendish Laboratory, Cambridge
Chapman & Hall Ltd., London
Edward Arnold & Co., Ltd., London
Institute of Metals, London
Institution of Gas Engineers, London
Iron and Steel Institute, London
Izvest. Akad. Nauk S.S.S.R.
Johnson, Matthey & Co. Ltd., London
John Wiley & Sons, Inc., New York
Journal of Scientific Instruments, London
Longmans, Green & Co. Ltd., London
Louis Cassier Co. Ltd., London
McGraw-Hill Book Co. Inc., New York
Melbourne University Press, Victoria, Australia
Oxford University Press, Oxford
Physical Review, Minnesota
Physical Society, London
Reinhold Publishing Corporation, New York
Royal Society, London
Society of Chemical Industry, London
Taylor & Francis Ltd., London
Zeitschrift für Physikalische Chemie, Leipzig

The assistance given by the following organizations is also gratefully acknowledged:

British Non-ferrous Metals Research Association
British Cast Iron Research Association
British Iron and Steel Research Association
British Ceramic Research Association

Extracts from the following British Standards are given by permission of the British Standards Institution, 2 Park Street, London, W.1, from whom official copies of the specifications may be obtained post free:

18 : 1956 (4s 6d)	1141 : 1943	1942: 1953 (2s 6d)
131 : Part 1: 1961 (6s)	1272–80: 1945 (4s)	2496: 1954 (2s. 6d)
131 : Part 2: 1959 (4s)	1400 : 1961 (20s)	2789: 1961 (5s)
210 : 1939 (2s 6d)	1452 : 1961 (5s)	2901: Part 1: 1957 (6s)
219 : 1959 (4s)	1453 : 1957 (6s)	2970: 1959 (8s 6d)
309 : 1958 (4s)	1475 : 1955 (7s)	2973: 1961 (6s)
310 : 1958 (4s)	1490 : 1955 (12s 6d)	3056: 1959 (5s)
321 : 1938	1501–6 : 1958 (25s)	3100: 1957 (7s 6d) containing
381C: 1944 (6s)	1507–8 : 1950 (10s 6d)	592 1462
410 : 1943 (6s)	1614 : 1949 (3s 6d)	1398 1463
467 : 1957 (5s)	1719 : 1951 (4s 6d)	1456 1617
786 : 1938	1758 : 1951 (2s. 6d)	1457 1630
991 : 1941	1826 : 1952 (8s 6d)	1458 1632
1004 : 1955 (3s 6d)	1845 : 1952 (3s)	1459 1648
1016 : Part 13: 1961 (7s 6d)	1902 : 1952 (12s 6d)	1461 1760
1017 : Part 2: 1960 (25s)	Add. 1: 1957 (3s)	3332: 1961 (5s)
1041 : 1943 (15s)	Add. 2: 1959 (3s)	2B21, 2B22 (2s 6d each)

EQUILIBRIUM DIAGRAMS

THE INTERCONVERSION OF ATOMIC AND WEIGHT PERCENTAGES IN BINARY SYSTEMS *

If W_x and A_x represent the weight and atomic percentages respectively of one component of a binary system, having atomic weight X, and if W_y, A_y and Y represent the corresponding quantities for the second component, then

$$W_y = 100 - W_x$$

and

$$A_y = 100 - A_x$$

The conversion from weight to atomic percentages, or conversely, may be made by means of the formulae

$$A_x = \frac{100 W_x}{W_x + \dfrac{X}{Y}(100 - W_x)} = \frac{100}{1 + \dfrac{X}{Y}\left(\dfrac{100}{W_x} - 1\right)}$$

$$W_x = \frac{100 A_x}{A_x + \dfrac{Y}{X}(100 - A_x)} = \frac{100}{1 + \dfrac{Y}{X}\left(\dfrac{100}{A_x} - 1\right)}$$

From these expressions the desired quantities may be obtained by means of tables of reciprocals or, if a lower order of accuracy is sufficient, by means of a slide rule. If many conversions are needed, use may be made of the rearranged equation

$$\frac{W_x}{100 - W_x} = \frac{X}{Y}\left(\frac{A_x}{100 - A_x}\right)$$

This equation lends itself readily to logarithmic computation. The left-hand side, and the term in brackets in the right-hand side, are both of the form $\dfrac{x}{100 - x}$, logarithms of which are given, for various values of x, in Table 1. Table 1 also gives the logarithms of the atomic weights, X and Y. If W_x and A_x are less than 1%, *i.e.* small compared with 100%, then the above equation reduces to the simpler form

$$W_x = \frac{X}{Y}A_x$$

The use of Table 1 may be made clear by two examples.
(a) What is the percentage of magnesium, by weight, in the compound Mg_2Si?
Here magnesium is the element about which information is sought, and it is therefore given the symbol and suffix x. Mg_2Si contains 66·67 at. %Mg. $X = 24\cdot32$, $A_x = 66\cdot67$.

From Table 1, $\log \dfrac{x}{100 - x}$ for $x = 66\cdot67$ ...   . 0·3011
Add $\log X = \log$ at. wt. Mg, from Table 1 ...   ... 1·3860

                                      1·6871
Subtract $\log Y = \log$ at. wt. Si ...   . 1·4486

                                        0·2385

This, from Table 1, is the $\log \dfrac{x}{100 - x}$ value for $x = 63\cdot40$, which is therefore W_x, the weight percentage required.
(b) What is the atomic percentage of copper in an aluminium–copper alloy containing 54·09% Cu by weight?

---

* From C. S. Smith, *Amer. Inst. Min. Met. Eng., Publ. No. 60, 1933.*

TABLE I.—VALUES OF LOG $\dfrac{x}{100-x}$

%	0	0·1	0·2	0·3	0·4	0·5	0·6	0·7	0·8	0·9	1	2	3	4	5	6	7	8	9
0	−∞	3̄·0004	·3019	·4784	·6038	·7012	·7808	·8482	·9066	·9582									
1	2̄·0044	·0462	·0844	·1196	·1522	·1827	·2111	·2379	·2632	·2871									
2	2̄·3098	·3314	·3521	·3718	·3908	·4089	·4264	·4433	·4595	·4752									
3	2̄·4903	·5050	·5193	·5331	·5465	·5595	·5722	·5846	·5966	·6083									
4	2̄·6198	·6310	·6419	·6526	·6630	·6732	·6832	·6930	·7026	·7120									
5	2̄·7212	·7303	·7392	·7479	·7565	·7649	·7732	·7814	·7894	·7973									
6	2̄·8050	·8127	·8202	·8276	·8349	·8421	·8492	·8562	·8631	·8699									
7	2̄·8766	·8832	·8898	·8962	·9026	·9089	·9151	·9213	·9274	·9334									
8	2̄·9393	·9452	·9510	·9567	·9624	·9680	·9736	·9791	·9845	·9899									
9	2̄·9952	**1̄·0005**	**·0057**	**·0109**	**·0160**	**·0211**	**·0261**	**·0311**	**·0360**	**·0409**									
10	1̄·0458	·0506	·0553	·0600	·0647	·0694	·0740	·0785	·0831	·0876									
11	1̄·0920	·0964	·1008	·1052	·1095	·1138	·1180	·1222	·1264	·1306									
12	1̄·1347	·1388	·1429	·1469	·1509	·1549	·1589	·1628	·1667	·1706									
13	1̄·1744	·1783	·1821	·1858	·1896	·1933	·1970	·2007	·2044	·2080									
14	1̄·2116	·2152	·2188	·2224	·2259	·2294	·2329	·2364	·2398	·2433	5	9	14	18	23	28	32	37	42
15	1̄·2467	·2501	·2534	·2568	·2602	·2635	·2668	·2701	·2734	·2766	4	9	13	17	21	26	30	34	38
16	1̄·2798	·2831	·2863	·2895	·2926	·2958	·2989	·3021	·3052	·3083	4	8	13	16	20	24	28	32	36
17	1̄·3114	·3145	·3175	·3205	·3236	·3266	·3296	·3326	·3356	·3385	4	8	12	16	20	24	28	32	33
18	1̄·3415	·3444	·3473	·3502	·3531	·3560	·3589	·3618	·3646	·3674	3	7	11	14	17	21	25	28	32
19	1̄·3703	·3731	·3759	·3787	·3815	·3842	·3870	·3898	·3925	·3952	3	7	10	13	16	20	23	26	30
20	1̄·3979	·4007	·4034	·4060	·4087	·4114	·4141	·4167	·4193	·4220	3	6	9	13	15	19	22	25	28
21	1̄·4246	·4272	·4298	·4342	·4350	·4376	·4401	·4427	·4453	·4478	3	6	9	12	14	18	21	24	27
22	1̄·4503	·4529	·4554	·4579	·4604	·4629	·4654	·4679	·4703	·4728	3	5	8	11	14	17	20	23	26
23	1̄·4752	·4777	·4801	·4826	·4850	·4874	·4898	·4922	·4946	·4970	3	5	8	11	13	16	19	21	25
24	1̄·4994	·5018	·5042	·5065	·5089	·5112	·5136	·5159	·5182	·5206	2	5	8	10	13	16	19	21	24
25	1̄·5229	·5252	·5275	·5298	·5321	·5344	·5367	·5389	·5412	·5435	2	5	8	10	13	15	18	21	23
26	1̄·5457	·5480	·5502	·5525	·5547	·5570	·5592	·5614	·5636	·5658	2	5	7	10	12	15	17	20	22
27	1̄·5680	·5702	·5724	·5746	·5768	·5790	·5812	·5833	·5855	·5877	2	4	7	9	12	14	17	19	22
28	1̄·5898	·5920	·5941	·5963	·5984	·6005	·6027	·6048	·6069	·6090	2	4	6	9	11	13	15	17	19
29	1̄·6111	·6132	·6154	·6175	·6196	·6216	·6237	·6258	·6279	·6300	2	4	6	8	11	13	15	17	19
30	1̄·6320	·6341	·6362	·6382	·6403	·6423	·6444	·6464	·6484	·6505	2	4	6	8	10	12	14	16	18
31	1̄·6525	·6545	·6566	·6586	·6606	·6626	·6646	·6666	·6686	·6706	2	4	6	8	10	12	14	16	18
32	1̄·6726	·6746	·6766	·6786	·6806	·6826	·6846	·6865	·6885	·6905	2	4	6	8	10	12	14	16	18
33	1̄·6924	·6944	·6964	·6983	·7003	·7022	·7042	·7061	·7081	·7100	2	4	6	8	10	12	14	16	18
34	1̄·7119	·7139	·7158	·7177	·7197	·7216	·7235	·7254	·7273	·7293	2	4	6	8	10	11	13	15	17
35	1̄·7312	·7331	·7350	·7369	·7388	·7407	·7426	·7445	·7463	·7482	2	4	6	8	9	11	13	15	17
36	1̄·7501	·7520	·7539	·7558	·7576	·7595	·7614	·7633	·7651	·7670	2	4	6	7	9	11	13	15	17
37	1̄·7689	·7707	·7726	·7744	·7763	·7781	·7800	·7819	·7837	·7856	2	4	5	7	9	11	13	15	17
38	1̄·7874	·7892	·7911	·7929	·7948	·7966	·7984	·8003	·8021	·8039	2	4	5	7	9	11	12	14	16
39	1̄·8057	·8076	·8094	·8112	·8130	·8148	·8167	·8185	·8203	·8221	2	4	5	7	9	11	12	14	16
40	1̄·8239	·8257	·8275	·8293	·8311	·8329	·8347	·8365	·8383	·8401	2	4	5	7	9	11	12	14	16
41	1̄·8419	·8437	·8455	·8473	·8491	·8509	·8527	·8545	·8563	·8580	2	4	5	7	9	11	12	14	16
42	1̄·8598	·8616	·8634	·8652	·8670	·8687	·8705	·8723	·8741	·8758	2	4	5	7	9	11	12	14	16
43	1̄·8776	·8794	·8811	·8829	·8847	·8864	·8882	·8900	·8917	·8935	2	4	5	7	9	11	12	14	16
44	1̄·8953	·8970	·8988	·9005	·9023	·9041	·9058	·9076	·9093	·9111	2	4	5	7	9	11	12	14	16
45	1̄·9129	·9146	·9164	·9181	·9199	·9216	·9234	·9251	·9269	·9286	2	3	5	7	9	10	12	14	16
46	1̄·9304	·9321	·9339	·9356	·9374	·9391	·9409	·9426	·9443	·9461	2	3	5	7	9	10	12	14	16
47	1̄·9478	·9496	·9513	·9531	·9548	·9565	·9583	·9600	·9618	·9635	2	3	5	7	9	10	12	14	16
48	1̄·9652	·9670	·9687	·9705	·9722	·9739	·9757	·9774	·9792	·9809	2	3	5	7	9	10	12	14	16
49	1̄·9826	·9844	·9861	·9878	·9896	·9913	·9931	·9948	·9965	·9983	2	3	5	7	9	10	12	14	16

Columns 1–9 give Mean Differences. For the upper rows: "Mean differences inaccurate, interpolate linearly."

Element	Z*	Log₁₀Z*
A	39·944	1·6015
Ac	277	2·4425
Ag	107·880	2·0329
Al	26·98	1·4310
Am	(243)	2·3856
As	74·91	1·8745
At	(210)	2·3222
Au	197·2	2·2949
B	10·82	1·0342
Ba	137·36	2·1379
Be	9·013	0·9549
Bi	209·00	2·3202
Bk	(245)	2·3892
Br	79·916	1·9026
C	12·010	1·0795
Ca	40·08	1·6029
Cd	112·41	2·0508
Ce	140·13	2·1465
Cf	(246)	2·3909
Cl	35·457	1·5497
Cm	(243)	2·3856
Co	58·94	1·7704
Cr	52·01	1·7161
Cs	132·91	2·1236
Cu	63·54	1·8031
Dy	162·46	2·2107
Er	167·2	2·2232
Eu	152·0	2·1818
F	19·00	1·2788
Fe	55·85	1·7470
Fr	(223)	2·3483
Ga	69·72	1·8434
Gd	156·9	2·1956
Ge	72·60	1·8609
H	1·008	0·0035
He	4·003	0·6024
Hf	178·6	2·2519
Hg	200·61	2·3024
Ho	164·94	2·2173
I	126·91	2·1035
In	114·76	2·0598
Ir	193·1	2·2858
K	39·100	1·5922
Kr	83·80	1·9232
La	138·92	2·1428
Li	6·940	0·8414
Lu	174·99	2·2430
Mg	24·32	1·3860
Mn	54·93	1·7398
Mo	95·95	1·9820

* Z = Atomic weight (1961 value, see page 46).

TABLE I.—continued

%	0	0·1	0·2	0·3	0·4	0·5	0·6	0·7	0·8	0·9	1	2	3	4	5	6	7	8	9
50	0·0000	·0017	·0035	·0052	·0070	·0087	·0104	·0122	·0139	·0156	2	3	5	7	9	10	12	14	16
51	0·0174	·0191	·0209	·0226	·0243	·0261	·0278	·0295	·0313	·0330	2	3	5	7	9	10	12	14	16
52	0·0348	·0365	·0382	·0400	·0417	·0435	·0452	·0470	·0487	·0504	2	3	5	7	9	10	12	14	16
53	0·0522	·0539	·0557	·0574	·0592	·0609	·0626	·0644	·0661	·0679	2	4	5	7	9	11	12	14	16
54	0·0696	·0714	·0731	·0749	·0766	·0784	·0801	·0819	·0836	·0854	2	4	5	7	9	11	12	14	16
55	0·0872	·0889	·0907	·0924	·0942	·0959	·0977	·0995	·1012	·1030	2	4	5	7	9	11	13	14	16
56	0·1047	·1065	·1083	·1100	·1118	·1136	·1153	·1171	·1189	·1206	2	4	5	7	9	11	13	14	16
57	0·1224	·1242	·1260	·1277	·1295	·1313	·1331	·1348	·1366	·1384	2	4	5	7	9	11	13	15	16
58	0·1402	·1420	·1437	·1455	·1473	·1491	·1509	·1527	·1545	·1563	2	4	5	7	9	11	13	15	16
59	0·1581	·1599	·1617	·1635	·1653	·1671	·1689	·1707	·1725	·1743	2	4	5	7	9	11	13	15	16
60	0·1761	·1779	·1797	·1815	·1833	·1852	·1870	·1888	·1906	·1924	2	4	5	7	9	11	13	15	16
61	0·1943	·1961	·1979	·1998	·2016	·2034	·2053	·2071	·2089	·2108	2	4	6	7	9	11	13	15	17
62	0·2126	·2145	·2163	·2182	·2200	·2219	·2237	·2256	·2274	·2293	2	4	6	7	9	11	13	15	17
63	0·2311	·2330	·2349	·2367	·2386	·2405	·2424	·2442	·2461	·2480	2	4	6	8	9	11	13	15	17
64	0·2499	·2518	·2537	·2555	·2574	·2593	·2612	·2631	·2650	·2669	2	4	6	8	9	11	13	15	17
65	0·2688	·2708	·2727	·2746	·2765	·2784	·2803	·2823	·2842	·2861	2	4	6	8	10	11	13	15	17
66	0·2881	·2900	·2919	·2939	·2958	·2978	·2997	·3017	·3036	·3056	2	4	6	8	10	12	14	16	18
67	0·3076	·3095	·3115	·3135	·3154	·3174	·3194	·3214	·3234	·3254	2	4	6	8	10	12	14	16	18
68	0·3274	·3294	·3314	·3334	·3354	·3374	·3394	·3414	·3434	·3455	2	4	6	8	10	12	14	16	18
69	0·3475	·3495	·3516	·3536	·3556	·3577	·3597	·3618	·3639	·3659	2	4	6	8	10	12	14	16	18
70	0·3680	·3701	·3721	·3742	·3763	·3784	·3805	·3826	·3847	·3868	2	4	6	8	11	13	15	17	19
71	0·3889	·3910	·3931	·3952	·3973	·3995	·4016	·4037	·4059	·4080	2	4	6	8	11	13	15	17	19
72	0·4102	·4123	·4145	·4167	·4188	·4210	·4232	·4254	·4276	·4298	2	4	7	9	11	13	15	18	20
73	0·4320	·4342	·4364	·4386	·4408	·4430	·4453	·4475	·4498	·4520	2	4	7	9	11	13	16	18	20
74	0·4543	·4565	·4588	·4611	·4633	·4656	·4679	·4702	·4725	·4748	2	5	7	9	11	14	16	18	20
75	0·4771	·4794	·4818	·4841	·4864	·4888	·4911	·4935	·4959	·4982	2	5	7	9	12	14	16	19	21
76	0·5006	·5030	·5054	·5078	·5102	·5126	·5150	·5174	·5199	·5223	2	5	7	10	12	15	17	19	22
77	0·5248	·5272	·5297	·5322	·5346	·5371	·5396	·5421	·5446	·5472	3	5	8	10	13	15	18	20	23
78	0·5497	·5522	·5548	·5573	·5599	·5624	·5650	·5676	·5702	·5728	3	5	8	11	13	16	19	21	24
79	0·5754	·5780	·5807	·5833	·5860	·5886	·5913	·5940	·5967	·5994	3	5	8	11	14	16	19	22	25
80	0·6021	·6048	·6075	·6103	·6130	·6158	·6185	·6213	·6241	·6269	3	5	8	11	14	17	20	23	25
81	0·6297	·6326	·6354	·6383	·6411	·6440	·6469	·6498	·6527	·6556	3	6	9	11	14	17	21	24	26
82	0·6581	·6615	·6645	·6674	·6704	·6734	·6764	·6795	·6825	·6856	3	6	9	12	15	18	22	25	27
83	0·6886	·6917	·6948	·6979	·7011	·7042	·7074	·7105	·7137	·7169	3	6	9	12	16	19	22	26	28
84	0·7202	·7234	·7267	·7299	·7332	·7365	·7398	·7432	·7466	·7499									
85	0·7533	·7567	·7602	·7636	·7671	·7706	·7741	·7776	·7812	·7848									
86	0·7884	·7920	·7956	·7993	·8030	·8067	·8104	·8142	·8180	·8218									
87	0·8256	·8294	·8333	·8372	·8411	·8451	·8491	·8531	·8571	·8612									
88	0·8653	·8694	·8736	·8778	·8820	·8862	·8905	·8948	·8992	·9036				Mean differences inaccurate, interpolate linearly					
89	0·9080	·9124	·9169	·9215	·9260	·9306	·9353	·9400	·9447	·9494									
90	0·9542	·9591	·9640	·9689	·9739	·9789	·9840	·9891	·9943	·9995									
91	1·0047	·0101	·0155	·0210	·0265	·0320	·0376	·0433	·0490	·0548									
92	1·0607	·0666	·0726	·0787	·0849	·0911	·0974	·1038	·1102	·1168									
93	1·1234	·1301	·1369	·1438	·1508	·1579	·1651	·1724	·1798	·1873									
94	1·1950	·2027	·2106	·2186	·2268	·2351	·2435	·2521	·2608	·2697									
95	1·2788	·2880	·2974	·3070	·3168	·3268	·3370	·3474	·3581	·3690									
96	1·3802	·3916	·4034	·4154	·4278	·4405	·4535	·4669	·4807	·4950									
97	1·5097	·5248	·5405	·5567	·5736	·5911	·6092	·6282	·6479	·6686									
98	1·6902	·7129	·7368	·7621	·7889	·8174	·8478	·8804	·9156	·9538									
99	1·9956	**2·0418**	**·0934**	**·1518**	**·2192**	**·2988**	**·3962**	**·5216**	**·6981**	**·9996**									

Element	Z*	Log₁₀ Z*
N	14·008	1·1464
Na	22·997	1·3617
Nb	92·91	1·9681
Nd	144·27	2·1592
Ne	20·183	1·3050
Ni	58·69	1·7686
Np	(237)	2·3748
O	16	1·2041
Os	190·2	2·2792
P	30·975	1·4910
Pa	231	2·3636
Pb	207·21	2·3164
Pd	106·7	2·0282
Pm	(145)	2·1614
Po	210	2·3222
Pr	140·92	2·1490
Pt	195·23	2·2905
Pu	(242)	2·3838
Ra	226·05	2·3542
Rb	85·48	1·9319
Re	186·31	2·2702
Rh	102·91	2·0125
Rn	222	2·3464
Ru	101·7	2·0073
S	32·066	1·5060
Sb	121·76	2·0855
Sc	44·96	1·6528
Se	78·96	1·8974
Si	28·09	1·4486
Sm	150·43	2·1773
Sn	118·70	2·0745
Sr	87·63	1·9427
Ta	180·88	2·2574
Tb	159·2	2·2019
Tc	(99)	1·9956
Te	127·61	2·1059
Th	232·12	2·3657
Ti	47·90	1·6803
Tl	204·39	2·3105
Tm	169·4	2·2289
U	238·07	2·3767
V	50·95	1·7071
W	183·92	2·2646
Xe	131·3	2·1183
Y	88·92	1·9490
Yb	173·04	2·2381
Zn	65·38	1·8154
Zr	91·22	1·9601

* Z = Atomic weight (1961 value, see page 46).

Here copper is assigned the symbol and suffix x. $W_x = 54 \cdot 09$.

$$\log \frac{x}{100 - x} \text{ for } x = 54 \cdot 09 \quad \ldots \quad \ldots \quad \ldots \quad \ldots \quad \ldots \quad 0 \cdot 0712$$

$$Add \log Y = \log \text{ at. wt. Al} \ldots \quad \ldots \quad \ldots \quad \ldots \quad \ldots \quad 1 \cdot 4310$$

$$1 \cdot 5022$$

$$Subtract \log X = \log \text{ at. wt. Cu} \quad \ldots \quad \ldots \quad \ldots \quad \ldots \quad 1 \cdot 8031$$

$$\bar{1} \cdot 6991$$

This, from Table 1, is the $\log \dfrac{x}{100 - x}$ value for $x = 33 \cdot 34$, which is therefore A_x, the atomic percentage of copper in the alloy.

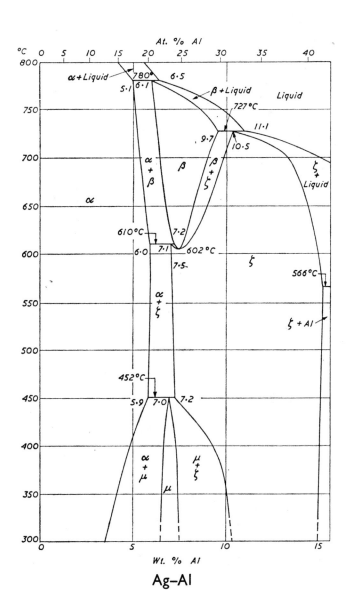

Ag–Al

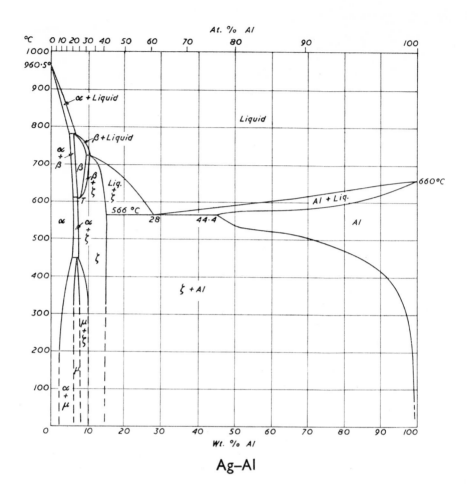

Ag–Al

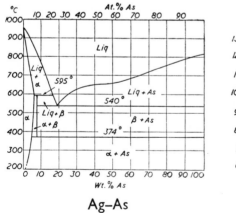

Ag–As

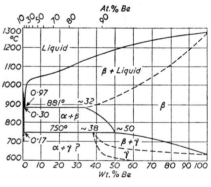

Ag–Be

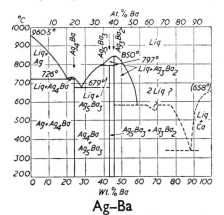

Ag–Ba

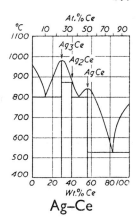

Ag–Ce

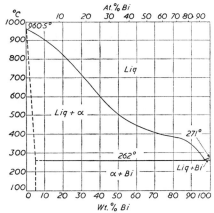

Ag–Bi

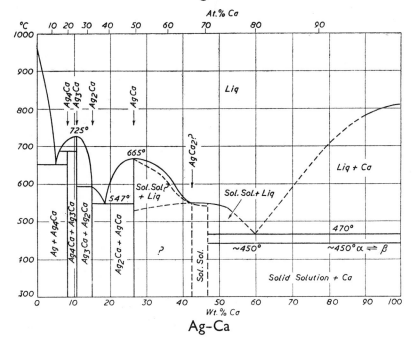

Ag-Ca

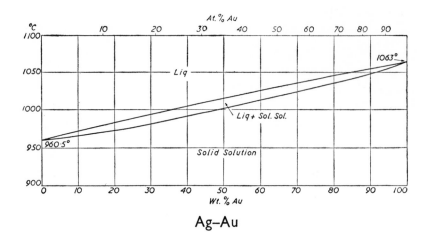

Ag–Au

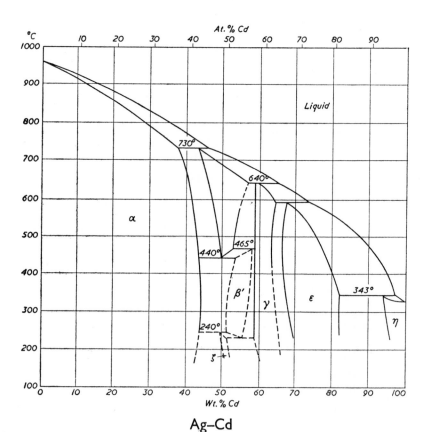

Ag–Cd

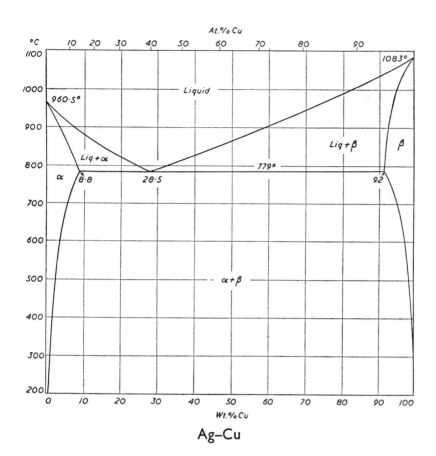

Ag–Cu

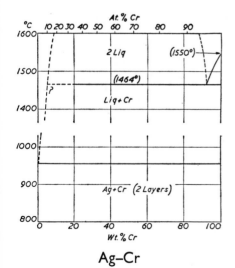

Ag–Cr

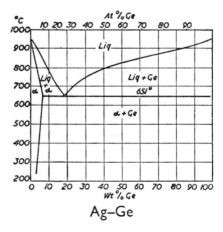

Ag–Ge

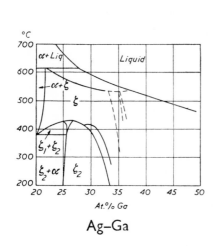

Ag–Ga

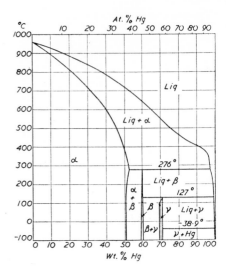

Ag–Hg

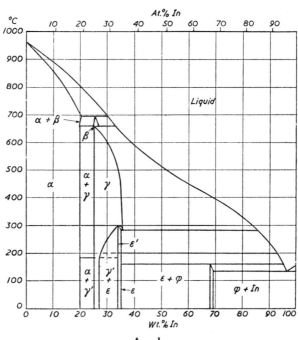

Ag–In

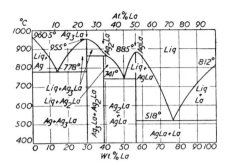

Ag–La

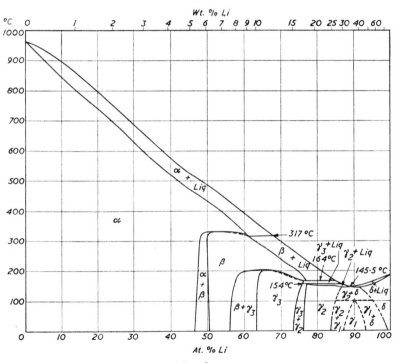

Ag–Li

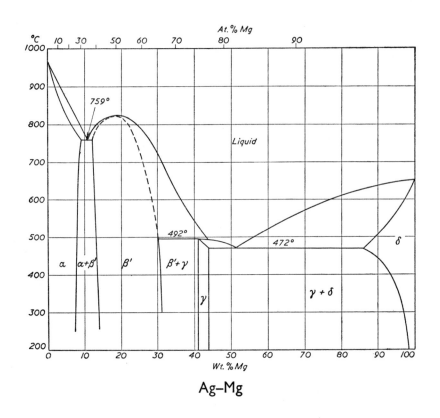

Ag–Mg

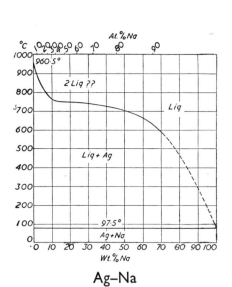

Ag–Na

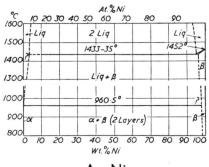

Ag–Ni

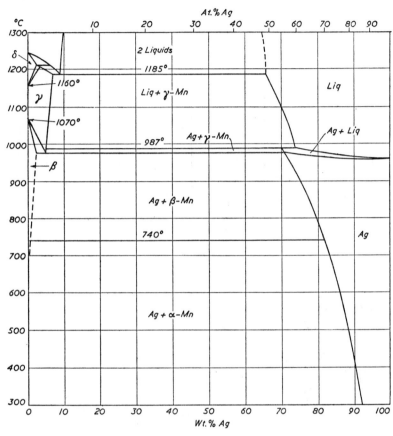

Ag–Mn

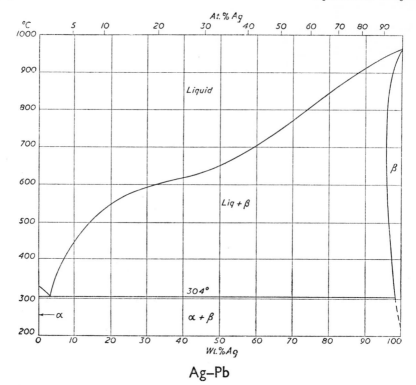

Ag–Pb

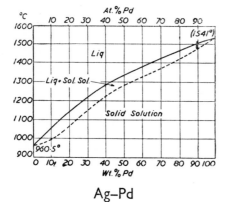

Ag–Pd

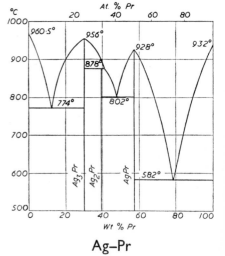

Ag–Pr

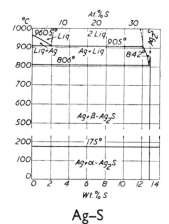

Ag–S

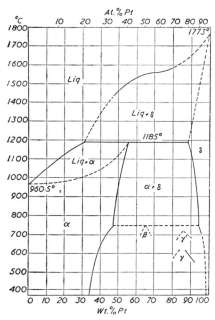

Ag–Pt

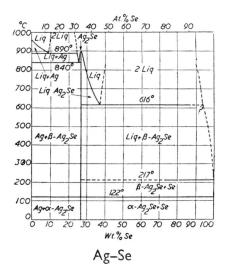

Ag–Se

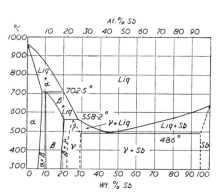

Ag–Sb

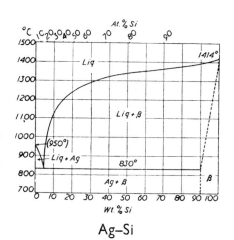

Ag–Si

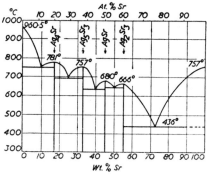

Ag–Sr

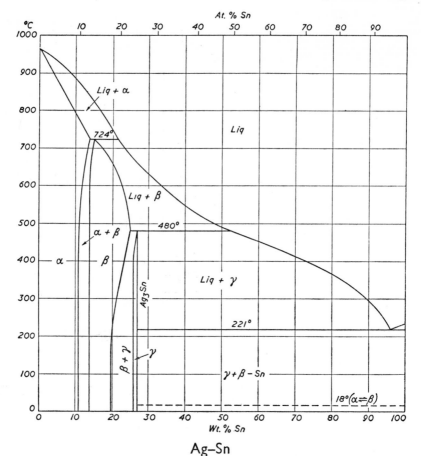

Ag–Sn

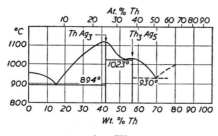

Ag–Th

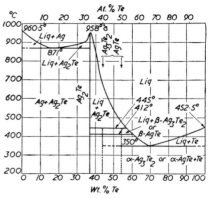

Ag–Te

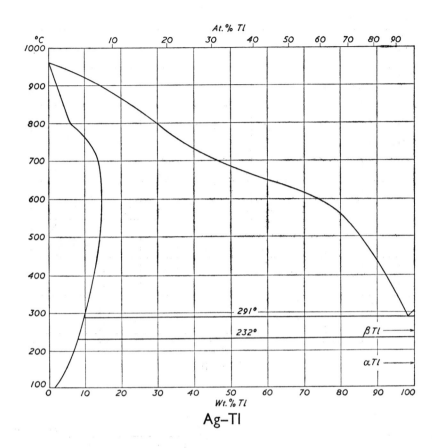

Ag–Tl

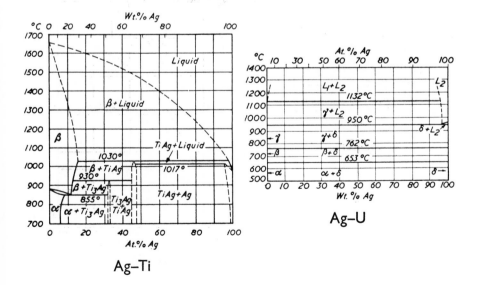

Ag–Ti

Ag–U

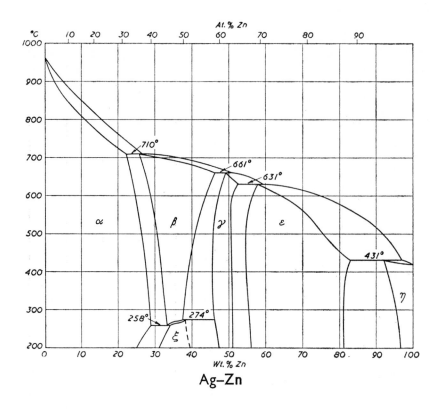

Ag–Zn

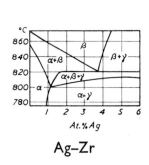

Ag–Zr

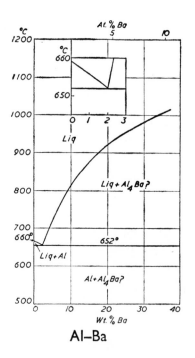

Al–Ba

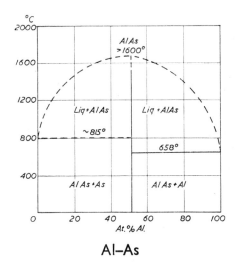

Al–As

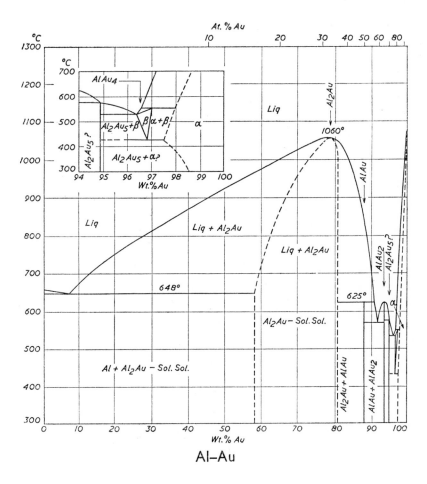

Al–Au

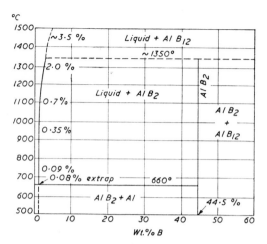

Al–B

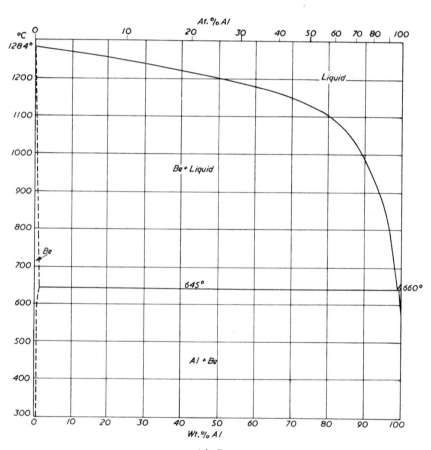

Al–Be

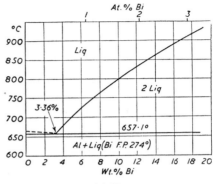

Al–Bi

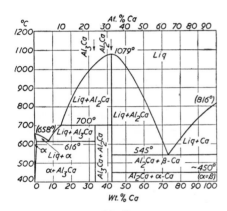

Al–Ca

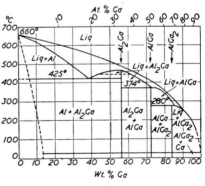

Al–Ga

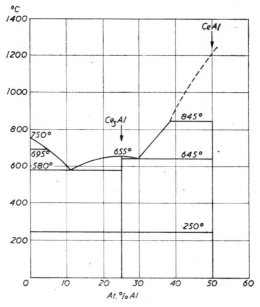

Al–Ce

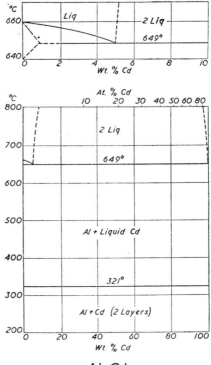

Al–Cd

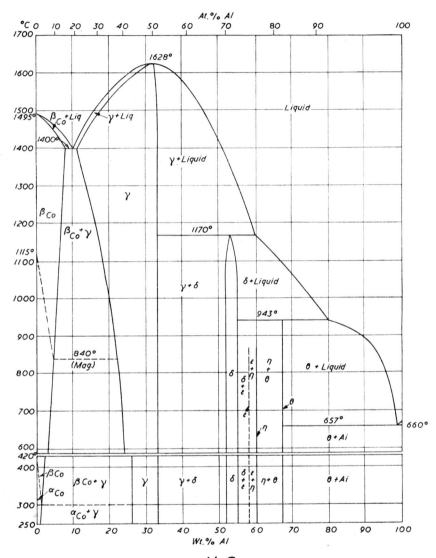

At. % Al

°C 0   10 20 30 40 50 60 70 80 90 100

1628°

β Co + Liq   γ + Liq

Liquid

1495°
1400°

γ + Liq

γ + Liquid

γ

β Co

β Co + γ

1170°

1115°
1100°

γ + δ   δ + Liquid

943°

840°
(Mag.)

δ   ε / η   η + θ   θ + Liquid
     δ+ε

θ

657°   660°

η

θ + Al

420°

β Co
α Co

β Co + γ   γ   γ + δ   δ δ+ε/η η + θ   θ + Al

α Co + γ

0   10   20   30   40   50   60   70   80   90   100

Wt. % Al

Al–Co

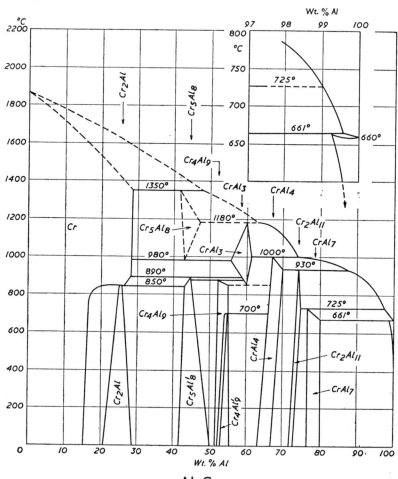

Al–Cr

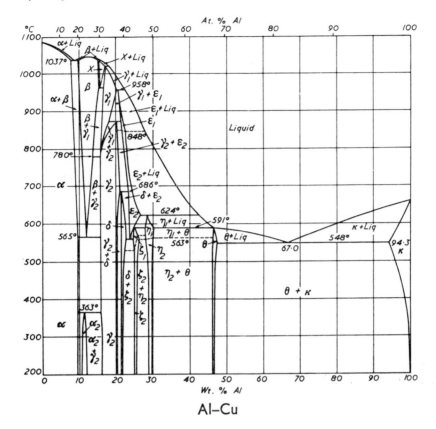

Al–Cu

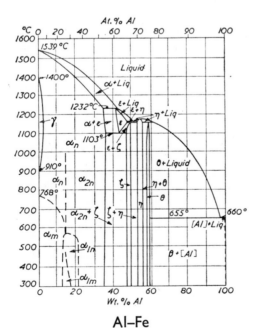

Al–Fe

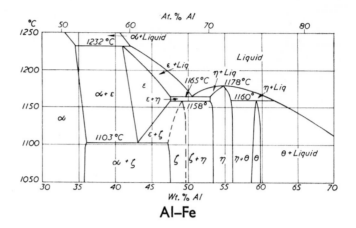

Al–Fe

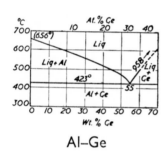

Al–Ge

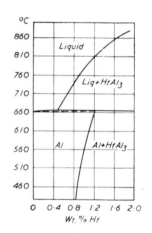

Al–Hf

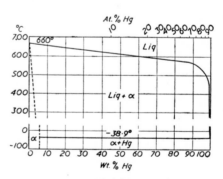

Al–Hg

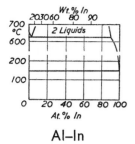

Al–In

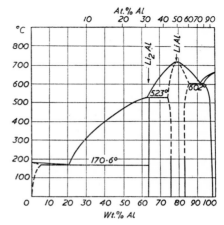

Al–Li

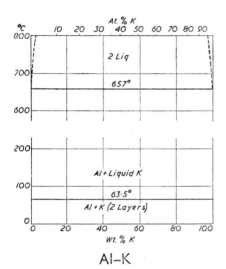

Al–K

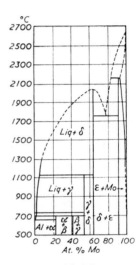

Al–Mo

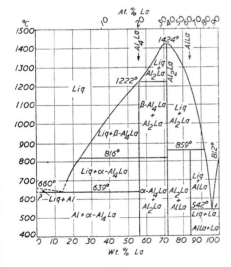

Al–La

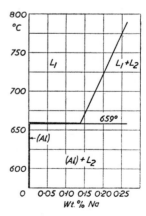

Al–Na

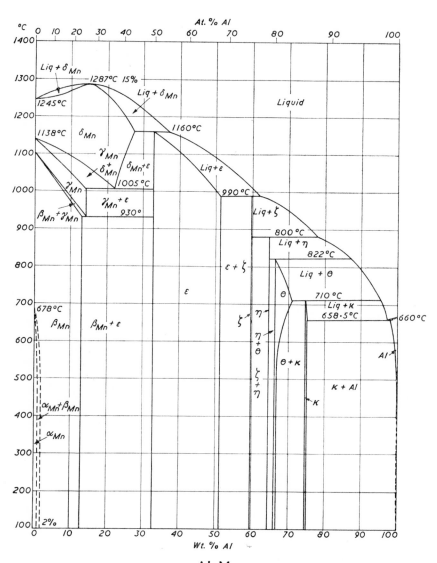

Al–Mn

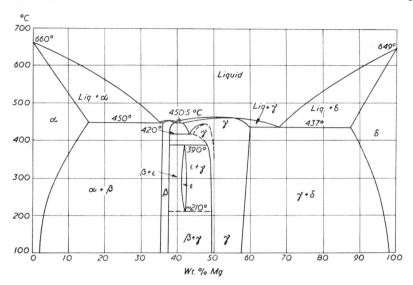

Al–Mg

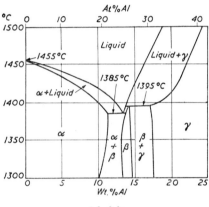

Al–Ni

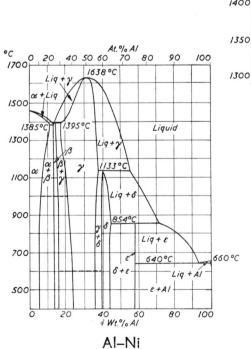

Al–Ni

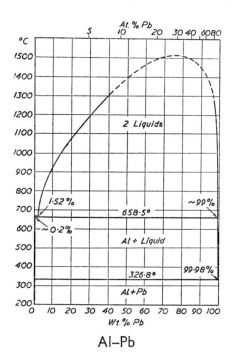

Al–Pb

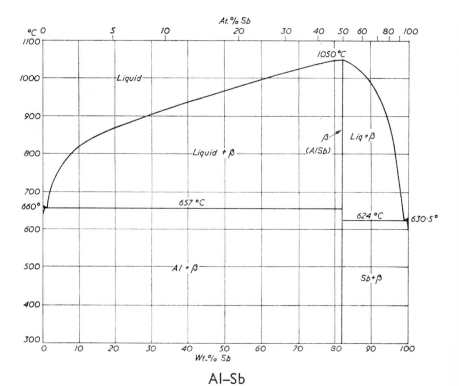

Al–Sb

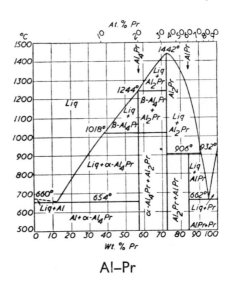

Al–Pr

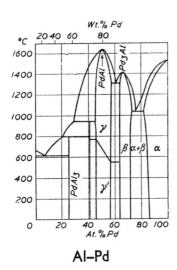

Al–Pd

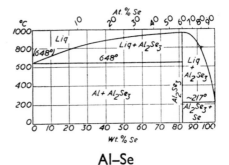

Al–Se

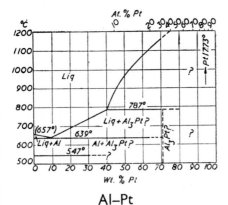

Al–Pt

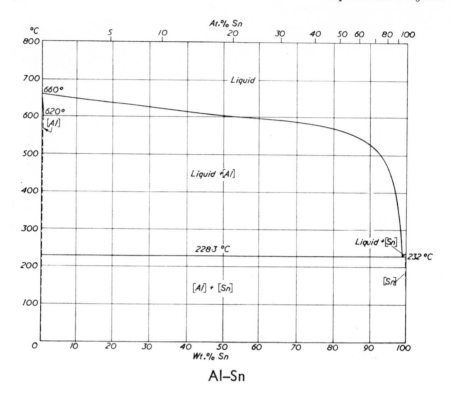

Al–Sn

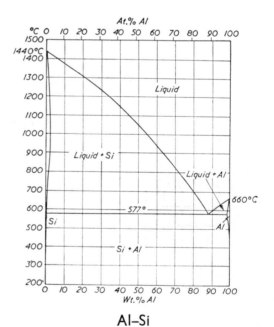

Al–Si

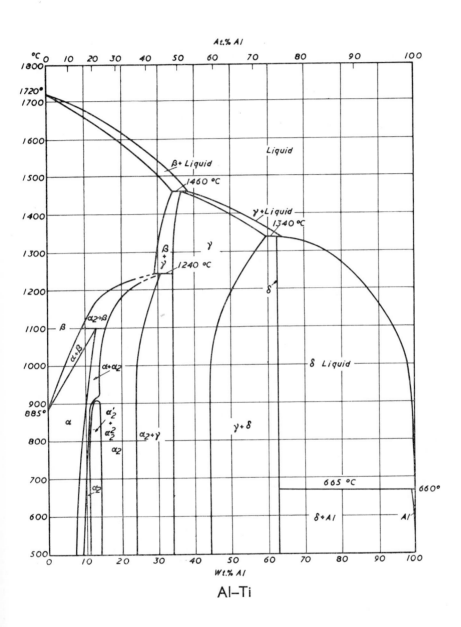

At.% Al

Al–Ti

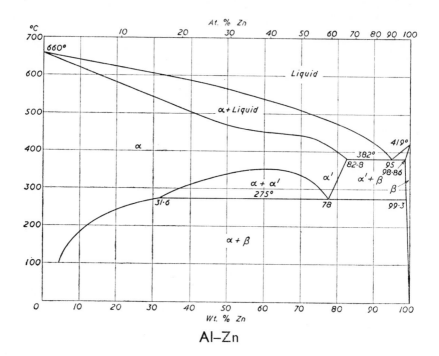

Al–Zn

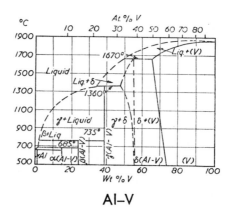

Al–V

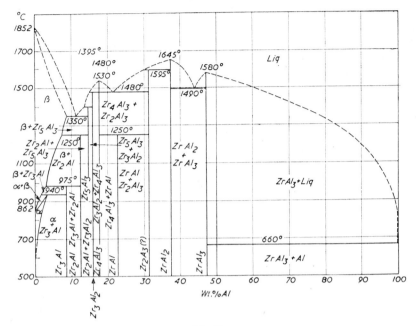

Al–Zr

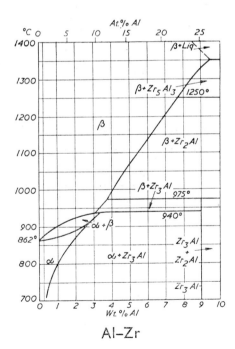

Al–Zr

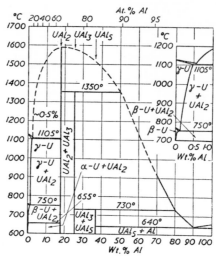

Al–U

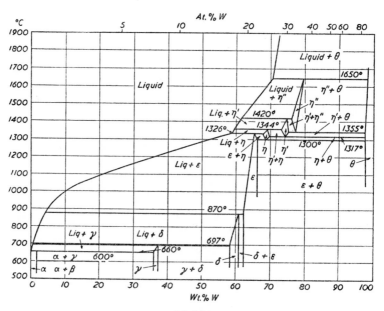

Al–W

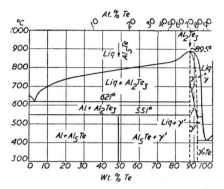

Al–Te

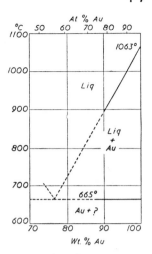

As–Au

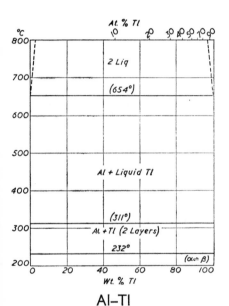

Al–Tl

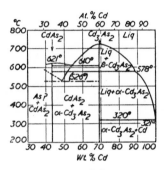

As–Cd

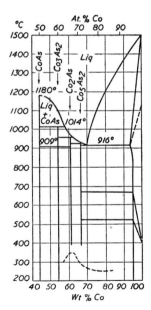

Al–Th

As–Co

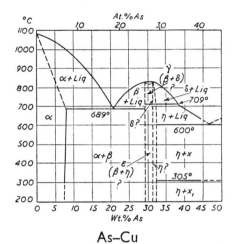

As–Cu

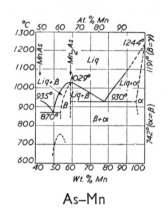

As–Mn

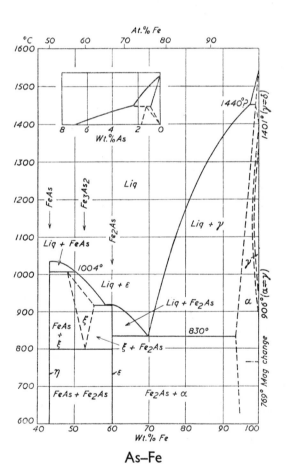

As–Fe

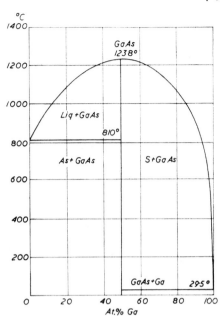

As–Ga

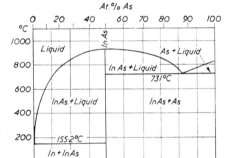

As–In

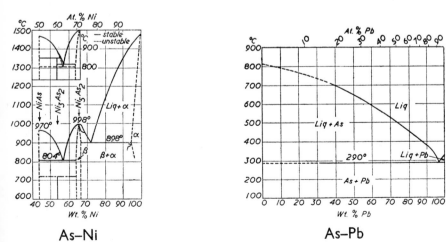

As–Ni

As–Pb

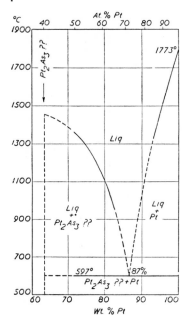

As–Pt

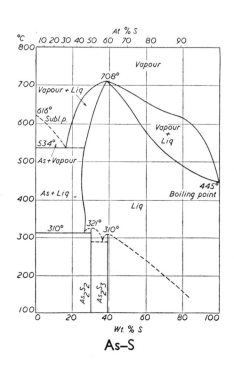

As–S

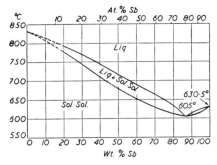

As–Sb

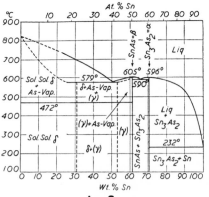

As–Sn

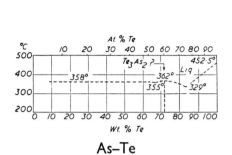

As–Te

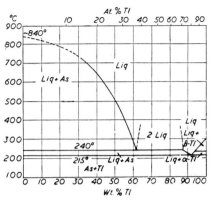

As–Tl

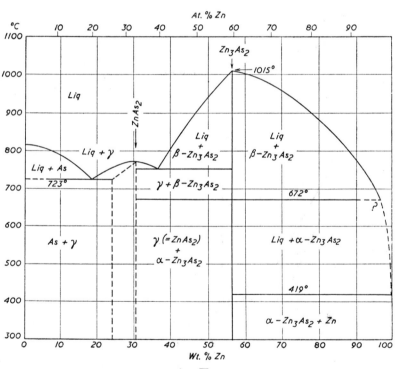

As–Zn

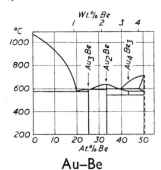

Au–Be

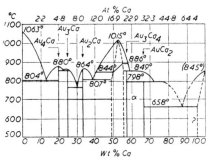

Au–Ca

Au–Bi

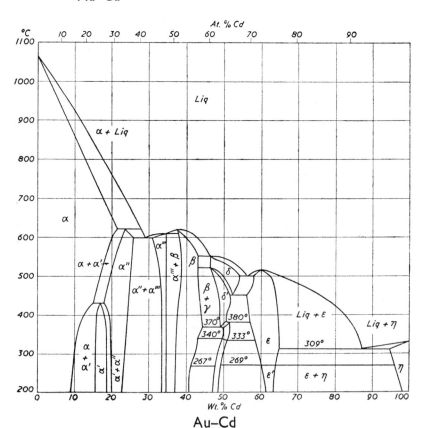

Au–Cd

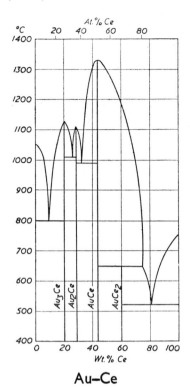

Au–Ce

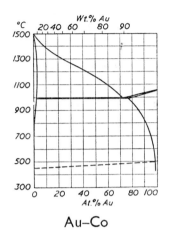

Au–Co

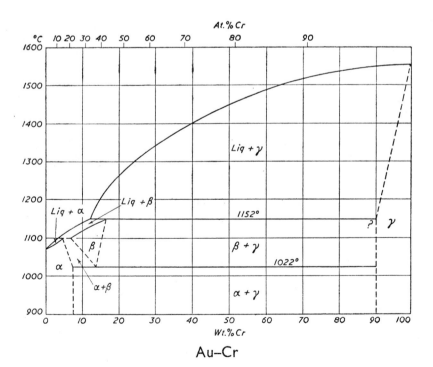

Au–Cr

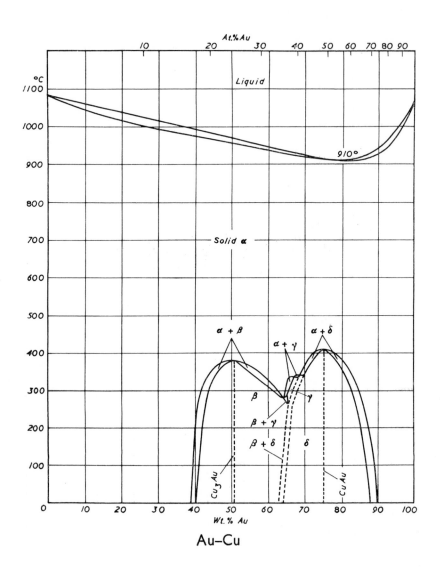

Au–Cu

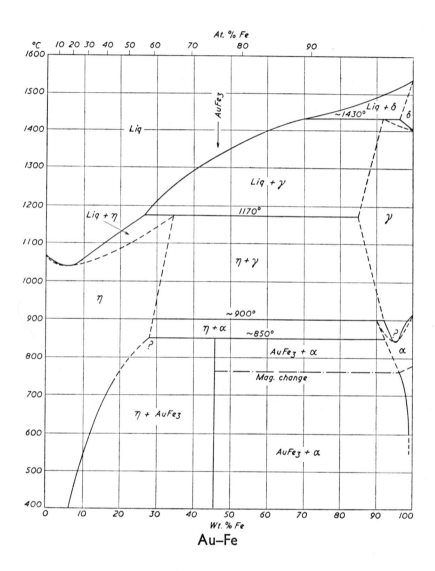

Au–Fe

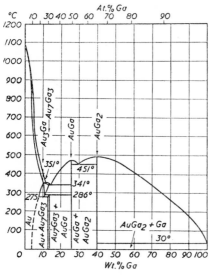

Au–Ga

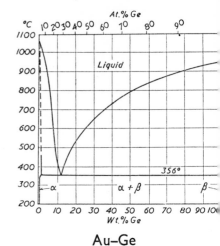

Au–Ge

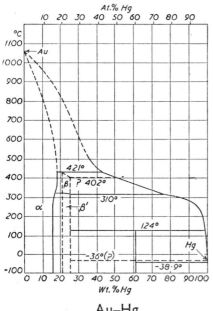

Au–Hg

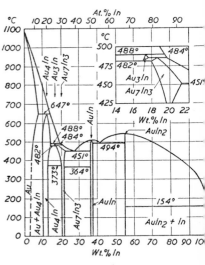

Au–In

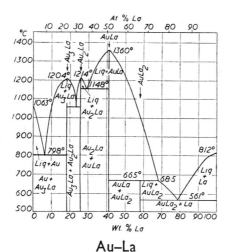

Au–La

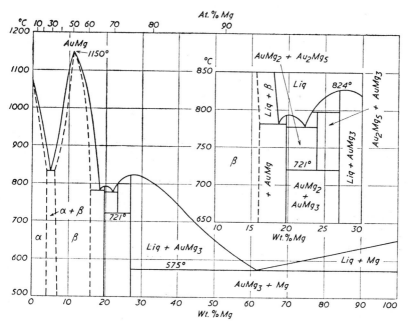

Au–Mg

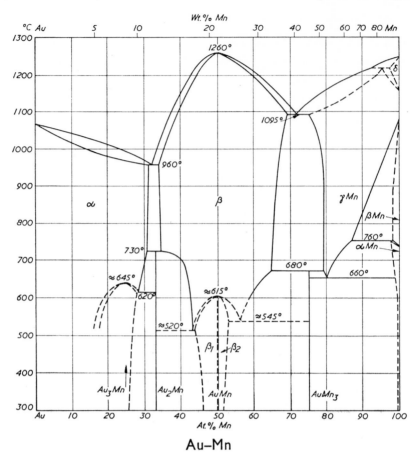

Au–Mn

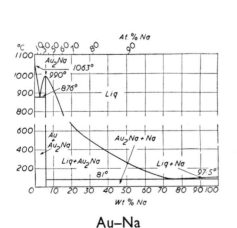

Au–Na

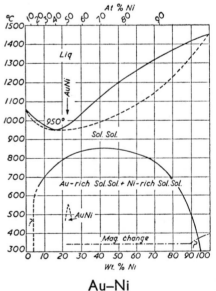

Au–Ni

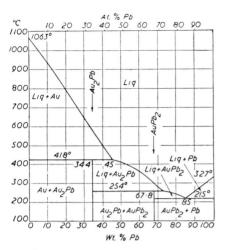

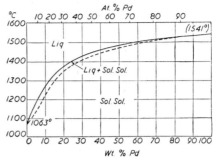

Au–Pd

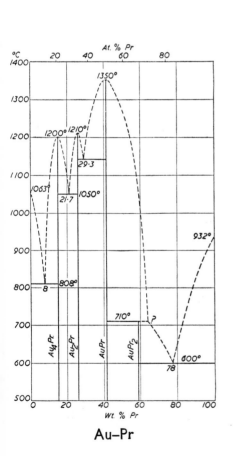

Au–Pb

Au–Pr

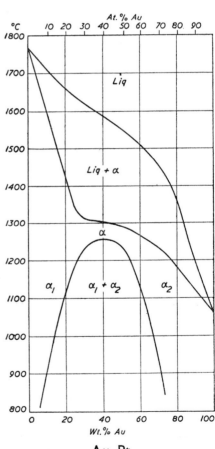

Au–Pt

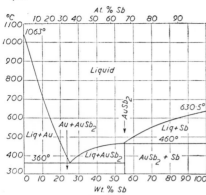

Au–Sb

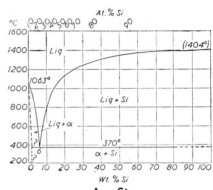

Au–Si

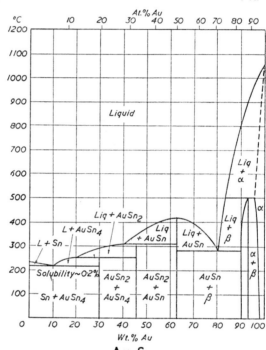

Au–Sn

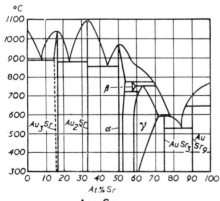

Au–Sr

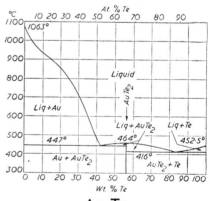

Au–Te

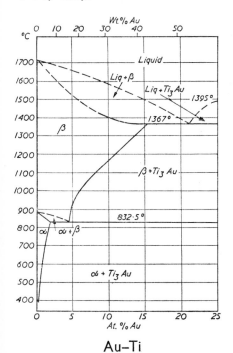

Au–Ti

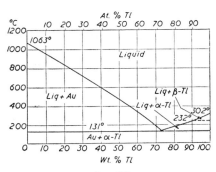

Au–Tl

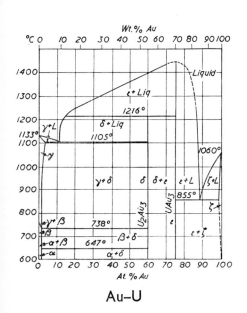

Au–U

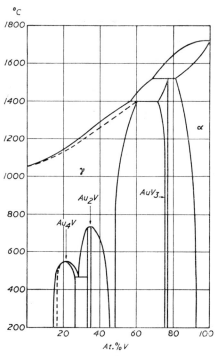

Au–V

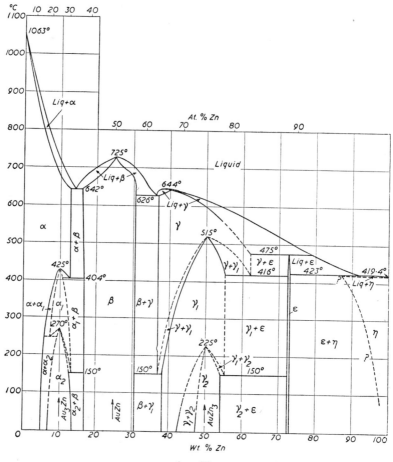

Au–Zn

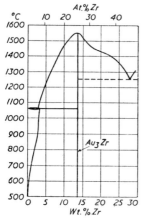

Au–Zr

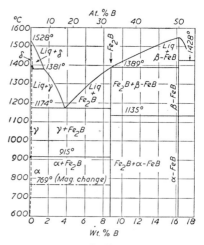

B–Fe

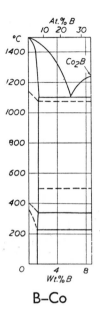

B–Co

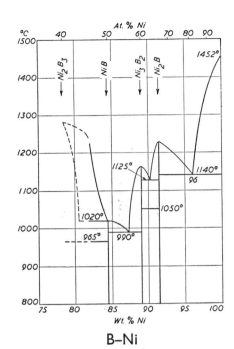

B–Ni

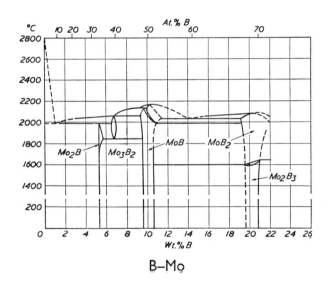

B–Mo

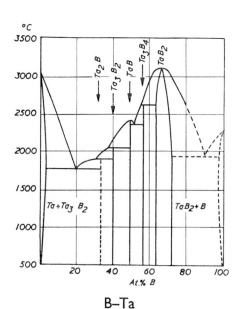

B–Ta

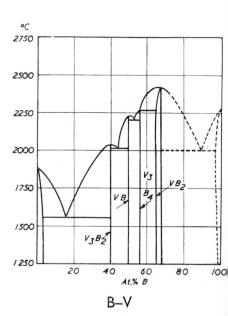

B–Nb

B–V

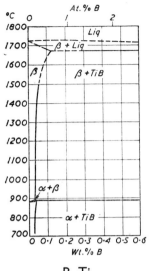

B–Ti

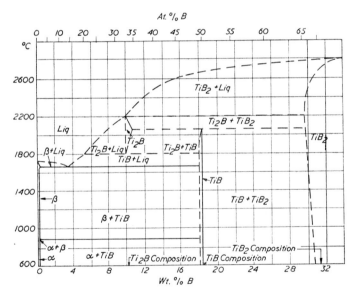

B–Ti

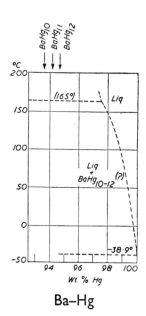

Ba–Hg

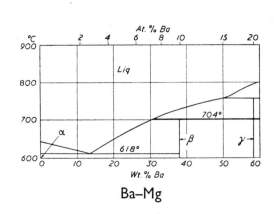

Ba–Mg

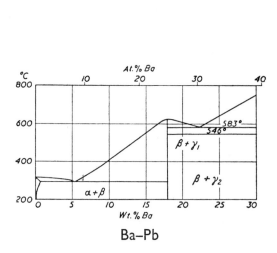

Ba–Pb

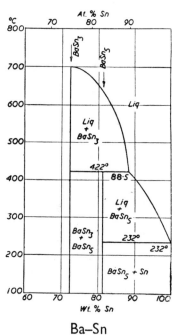

Ba–Sn

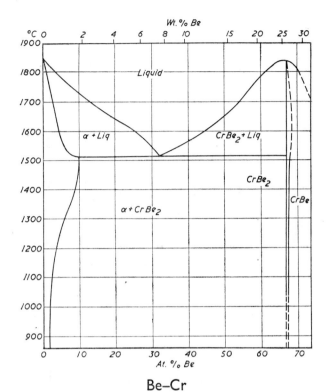

Be–Cr

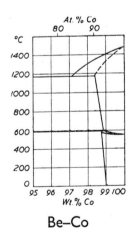

Be–Co

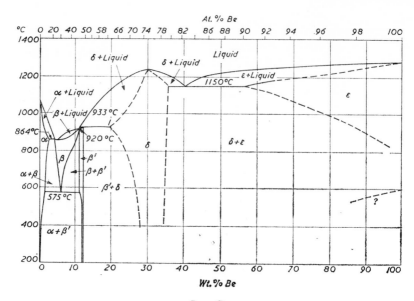

Be–Cu

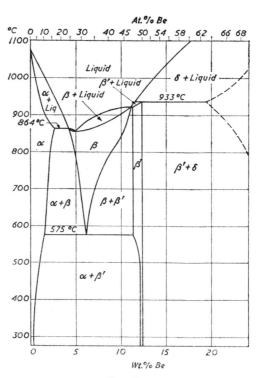

Be–Cu

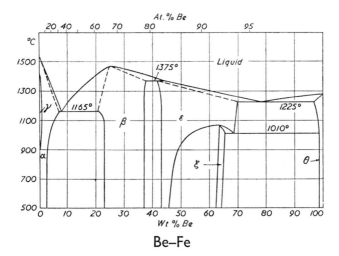

Be–Fe

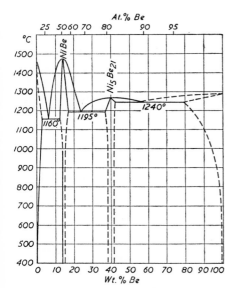

Be–Ni

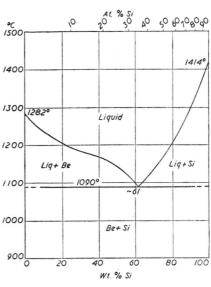

Be–Si

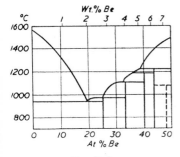

Be–Pd

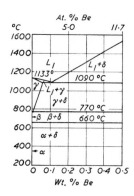

Be–U

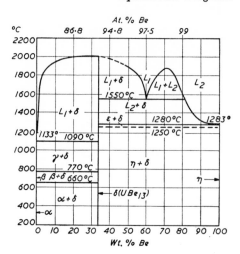

Be–U

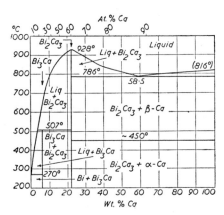

Bi–Ca

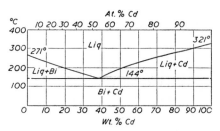

Bi–Cd

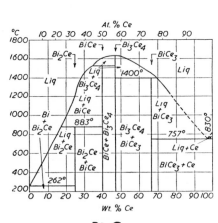

Bi–Ce

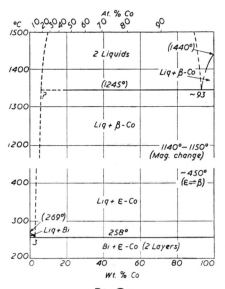

Bi–Co

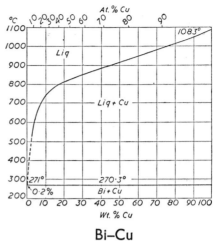

Bi–Cr

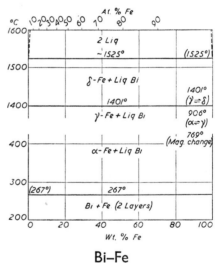

Bi–Cu

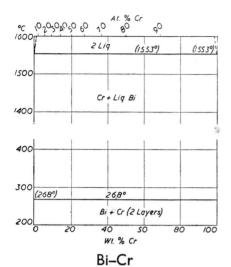

Bi–Ga

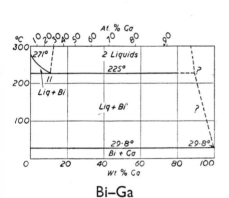

Bi–Fe

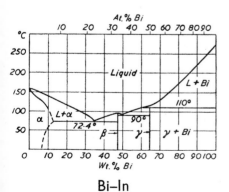

Bi–In

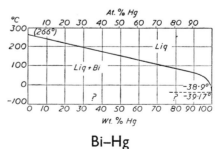

Bi–Hg

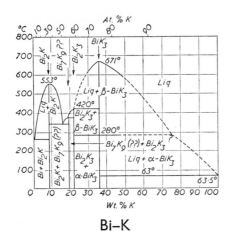

Bi–K

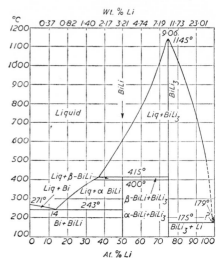

Bi–Li

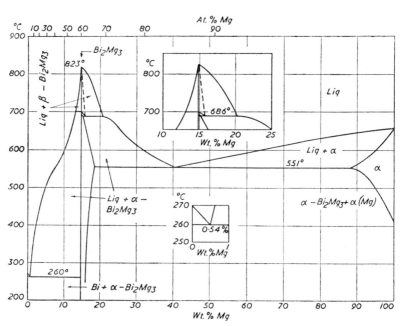

Bi–Mg

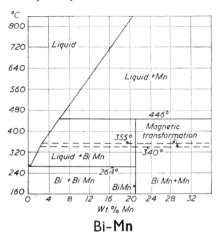

Bi-Mn

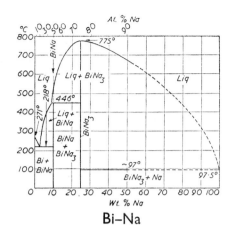

Bi-Na

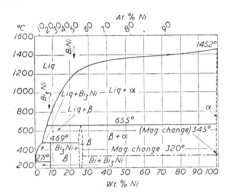

Bi–Ni

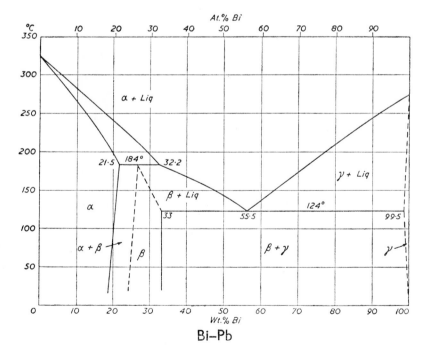

Bi–Pb

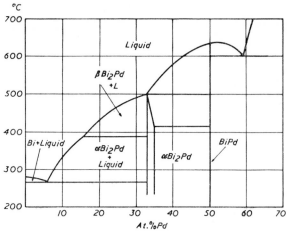

Bi–Pd

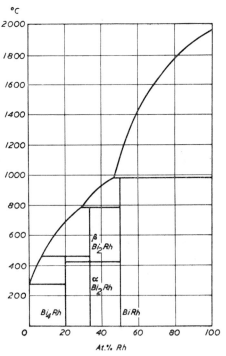

Bi–Rh

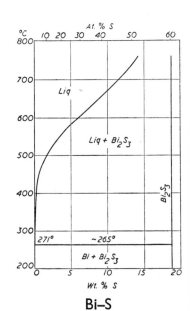

Bi–S

Bi–Sb

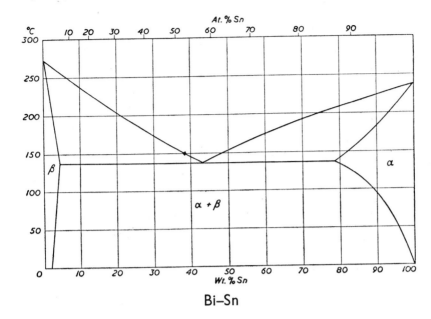

Bi–Sn

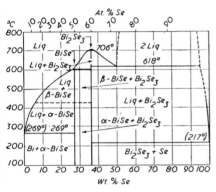

Bi–Se

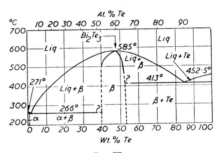

Bi–Te

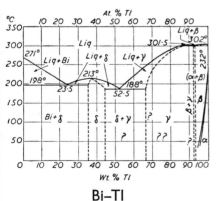

Bi–Tl

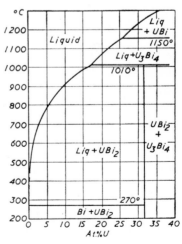

Bi–U

D (II)

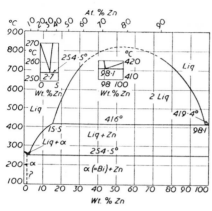

Bi–Zn

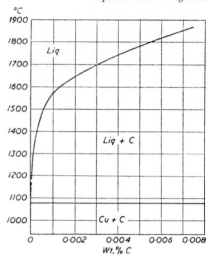

C–Cu

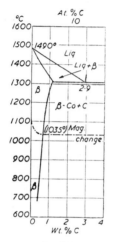

C–Co

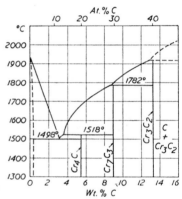

C–Cr

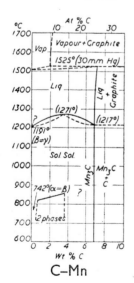

C–Mn

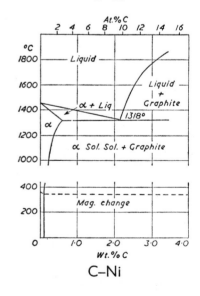

C–Ni

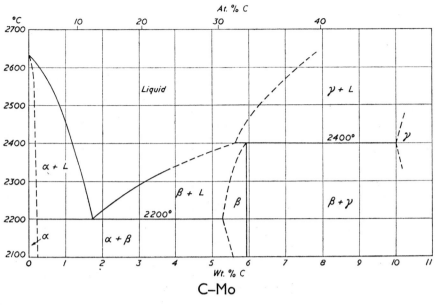

At. % C

°C

Liquid

$\gamma + L$

2400°

γ

$\alpha + L$

$\beta + L$

β

$\beta + \gamma$

2200°

α

$\alpha + \beta$

Wt. % C

C–Mo

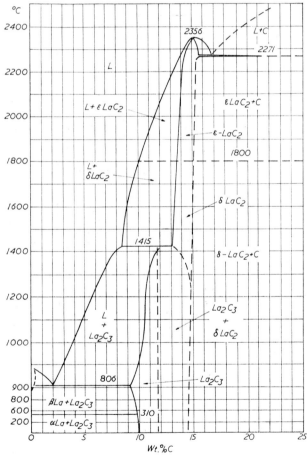

°C

2356

L+C

2271

L

$L + \varepsilon\,LaC_2$

$\varepsilon LaC_2 + C$

$\varepsilon\text{-}LaC_2$

1800

$L + \delta LaC_2$

$\delta\,LaC_2$

1415

$\delta\text{-}LaC_2 + C$

$L + La_2C_3$

$La_2C_3 + \delta\,LaC_2$

806

La_2C_3

900
800
600
200

$\beta La + La_2C_3$

310

$\alpha La + La_2C_3$

Wt. % C

C–La

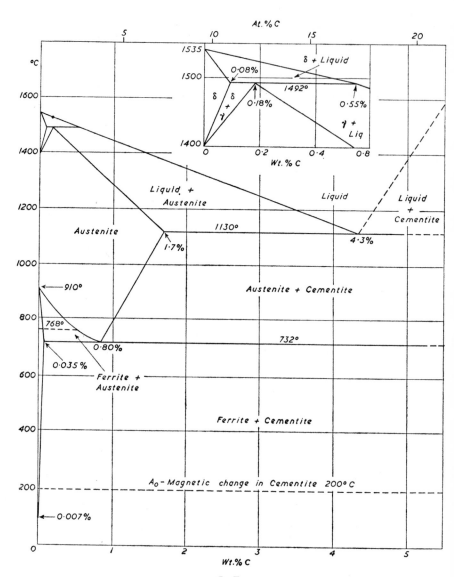

C–Fe

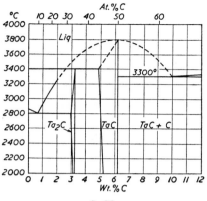

C–Ta

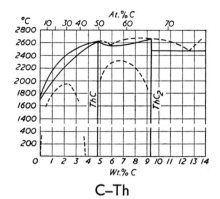

C–Th

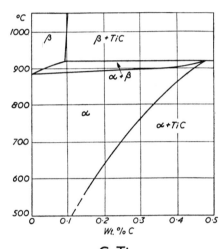

C–Ti

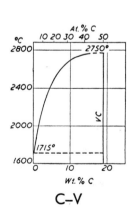

C–V

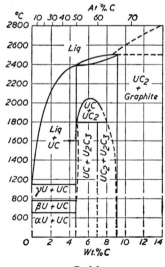

C–U

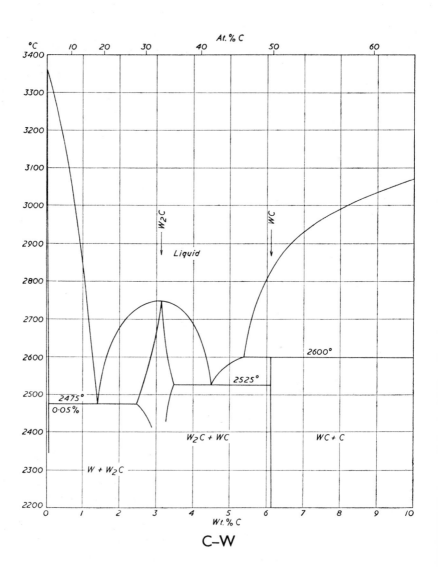

C–W

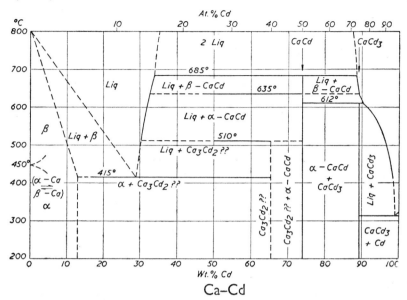

Ca–Cd

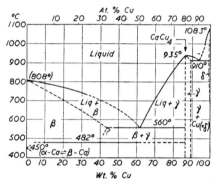

Ca–Cu

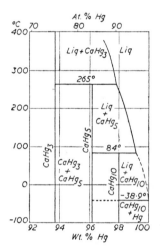

Ca–Hg

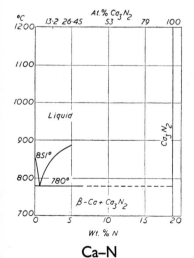

Ca–N

Ca–Mg

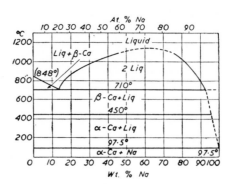

Ca–Na

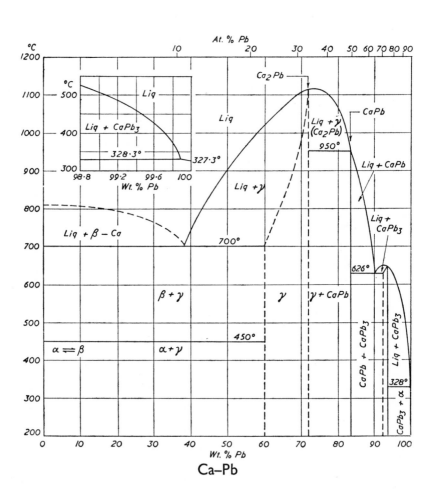

Ca–Pb

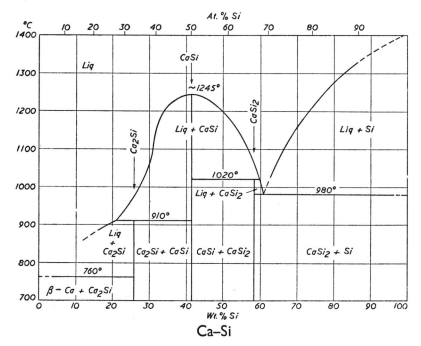

Ca–Si

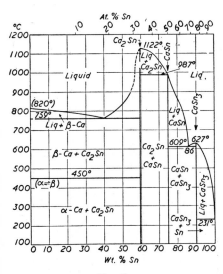

Ca–Sn

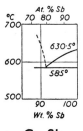

Ca–Sb

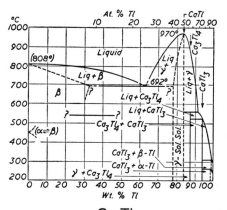

Ca–Tl

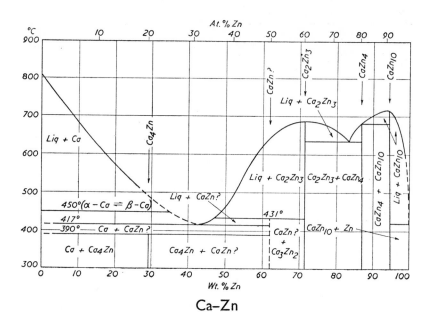

Ca–Zn

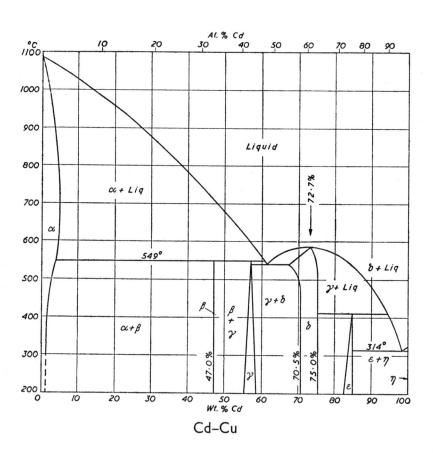

Cd–Cu

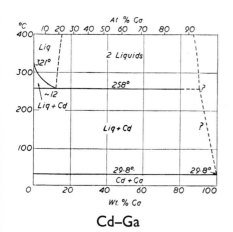

Cd–Ga

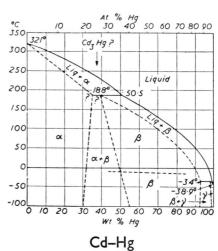

Cd–Hg

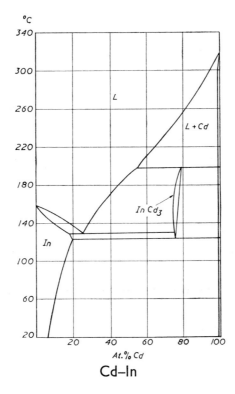

Cd–In

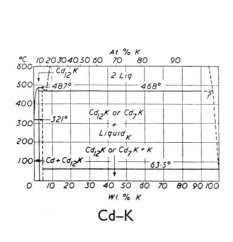

Cd–K

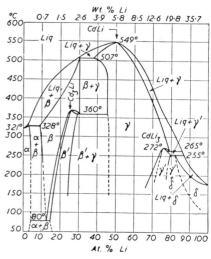

Cd–Li

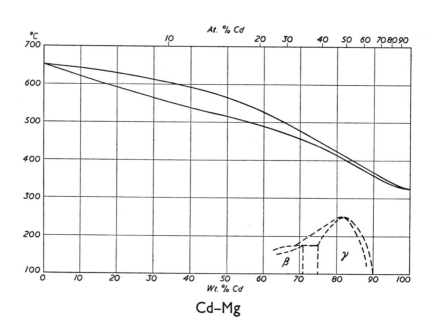

Cd–Mg

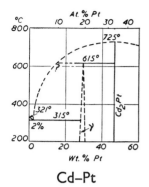

Cd–Pt

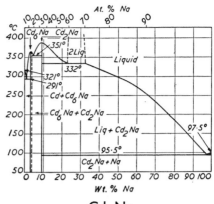

Cd–Na

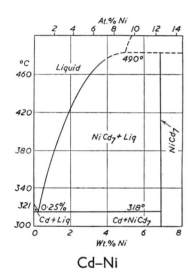

Cd–Ni

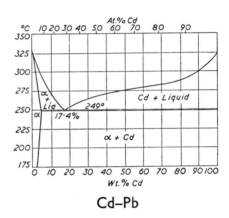

Cd–Pb

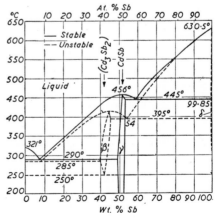

Cd–Te

Cd–Sb

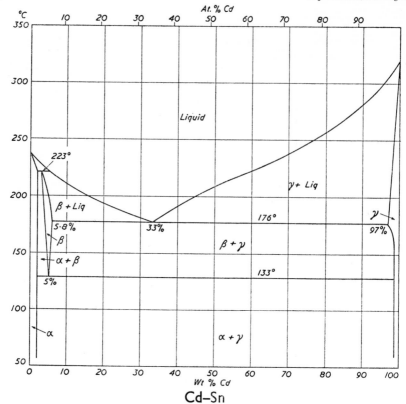

Cd–Sn

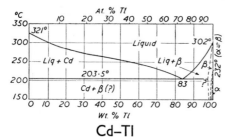

Cd–Tl

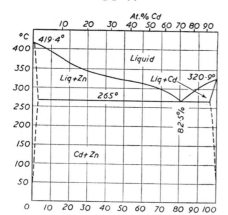

Cd–Zn

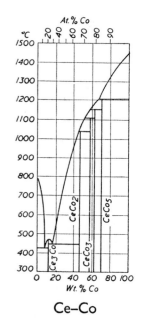

Ce–Co

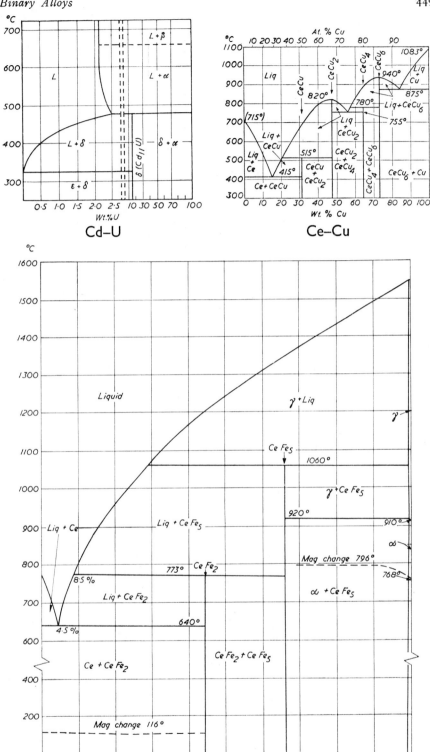

Cd–U

Ce–Cu

Ce–Fe

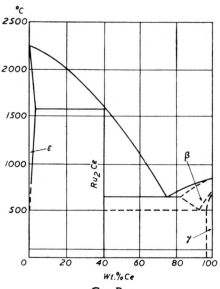

Ce–Ru

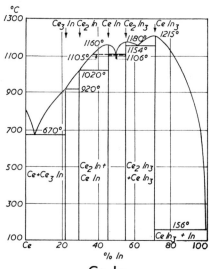

Ce–In

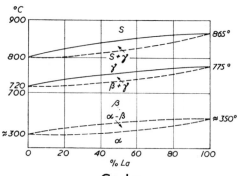

Ce–La

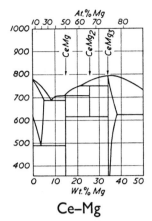

Ce–Mg

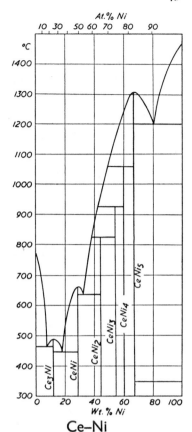

Ce–Ni

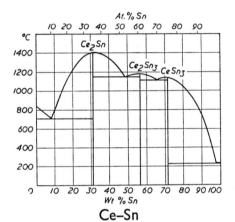

Ce–Sn

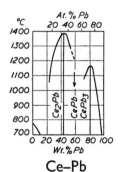

Ce–Pb

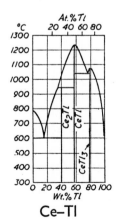

Ce–Tl

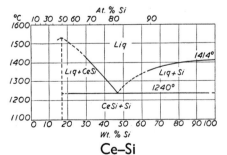

Ce–Si

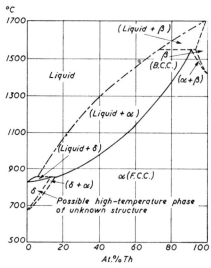

Ce–Th

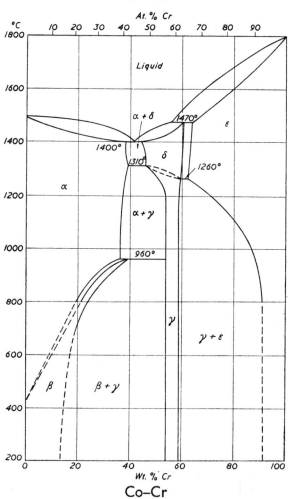

Co–Cr

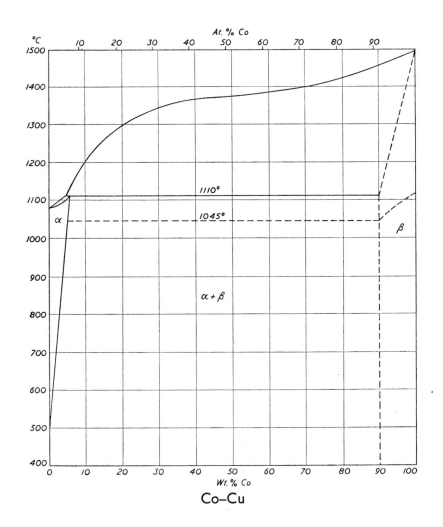

Co–Cu

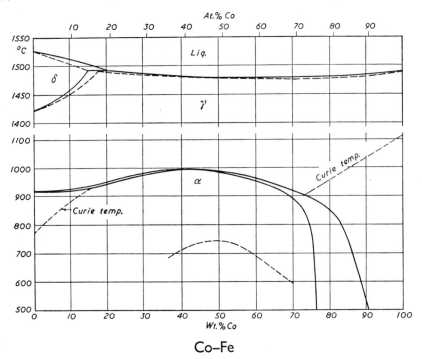

Co–Fe

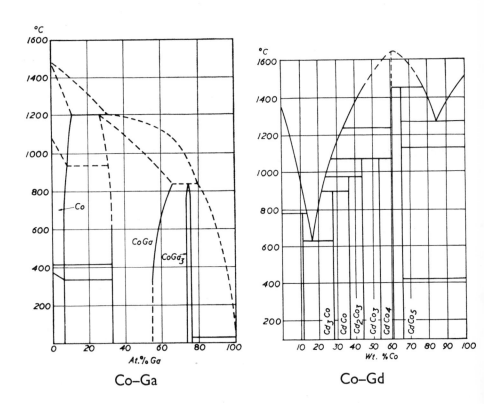

Co–Ga

Co–Gd

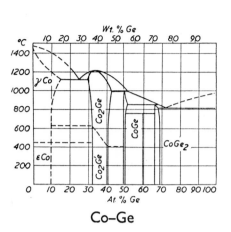

Co–Ge

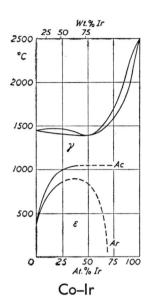

Co–Ir

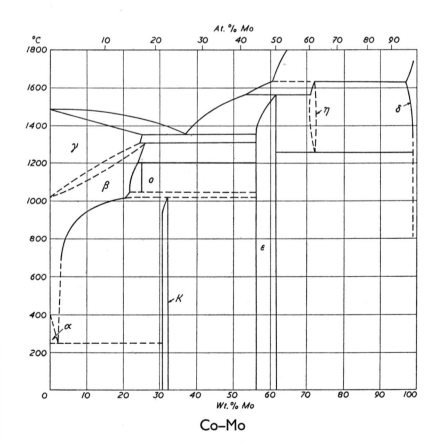

Co–Mo

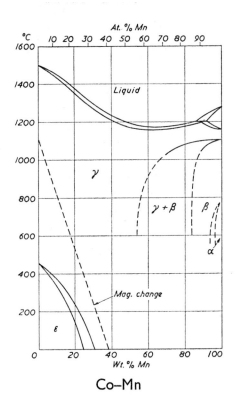

Co–Mn

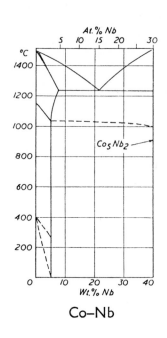

Co–Nb

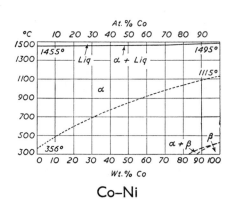

Co–Ni

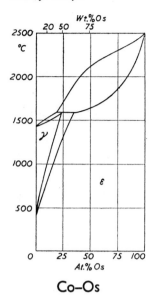

Co–Os

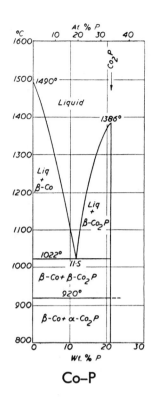

Co–P

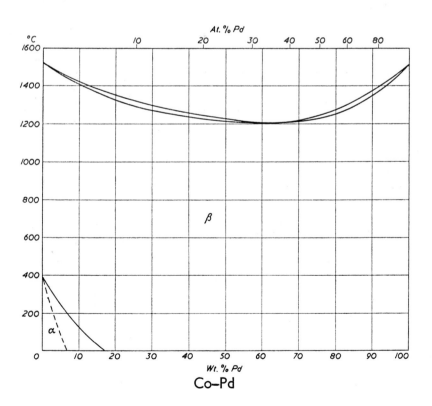

Co–Pd

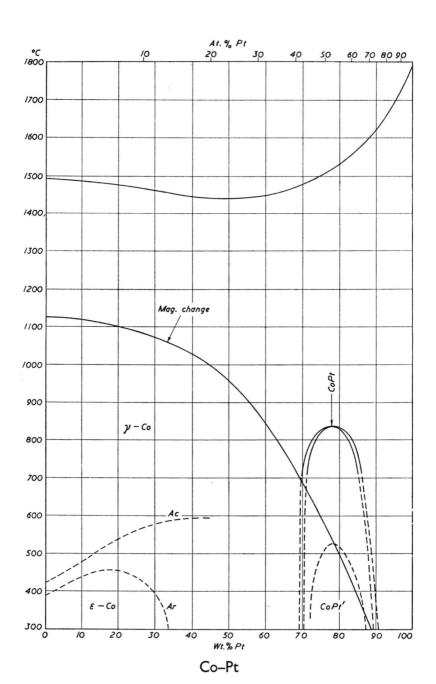

Co–Pt

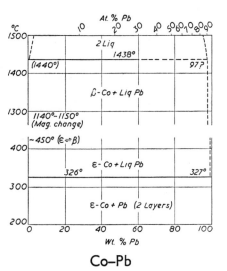

Co–Pb

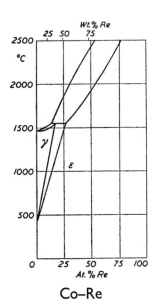

Co–Re

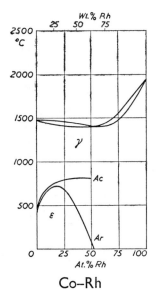

Co–Rh

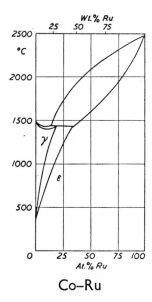

Co–Ru

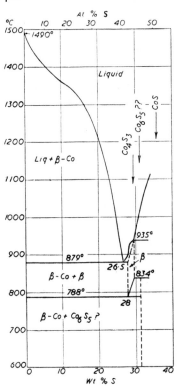

Co–S

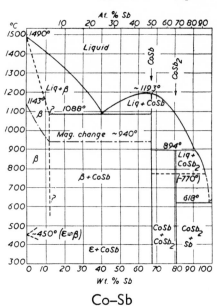

Co–Sb

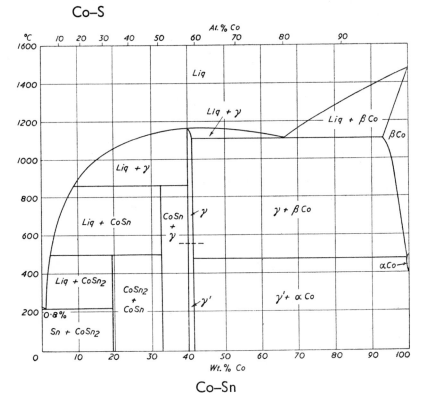

Co–Sn

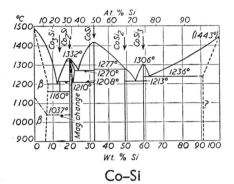

Co–Si

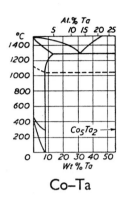

Co–Ta

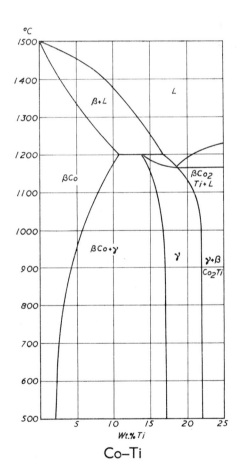

Co–Ti

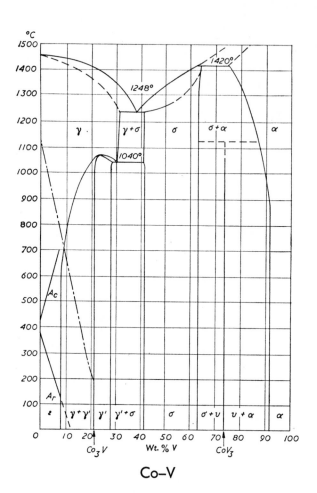

Co–V

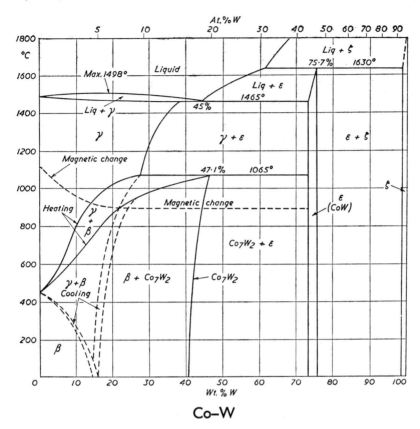

Co–W

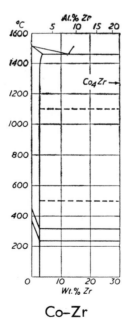

Co–Zr

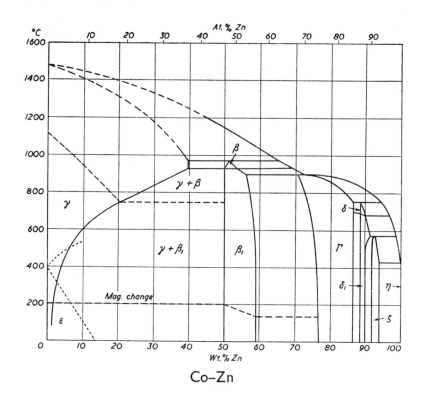

Co–Zn

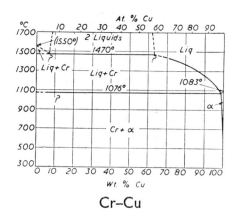

Cr–Cu

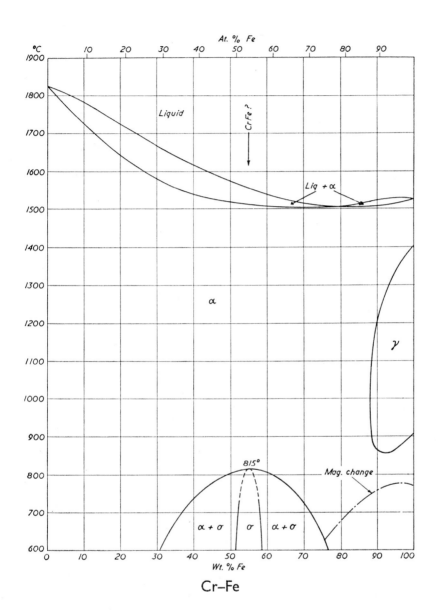

Cr–Fe

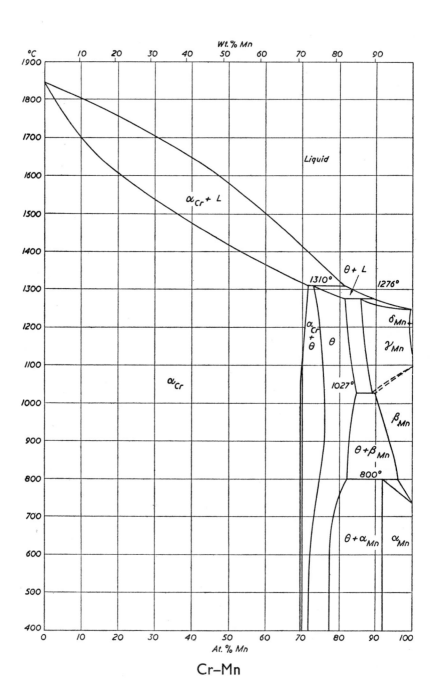

Cr–Mn

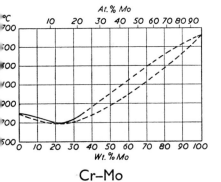

Cr–Mo

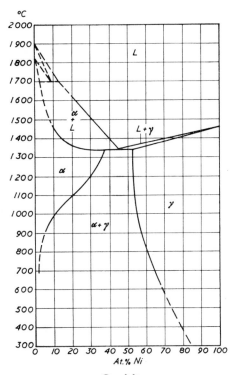

Cr–Ni

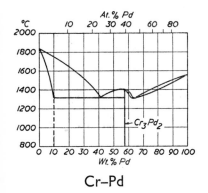

Cr–Pd

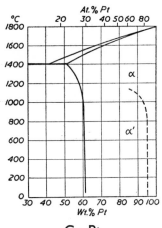

Cr–Pt

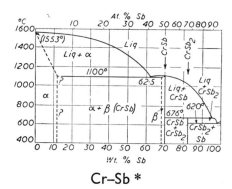

Cr–Sb *

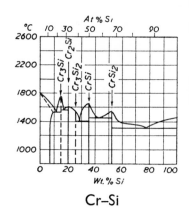

Cr–Si

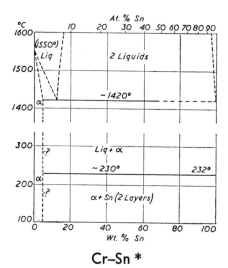

Cr–Sn *

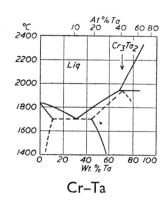

Cr–Ta

---

* Diagram obviously unreliable, but no more recent determinations available.

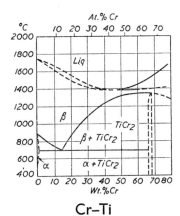

Cr–Ti

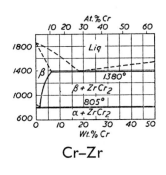

Cr–Zr

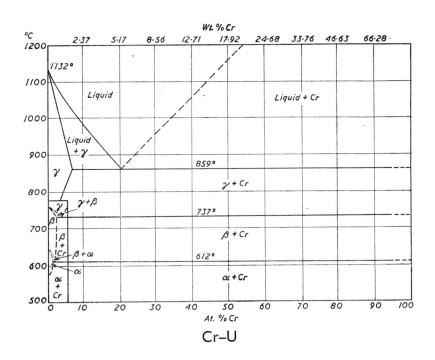

Cr–U

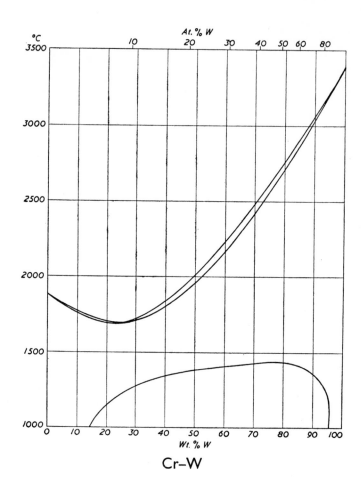

Cr–W

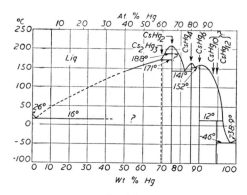

Cs–Hg

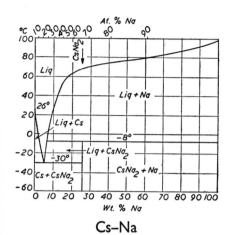

Cs–Na

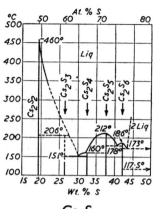

Cs–S

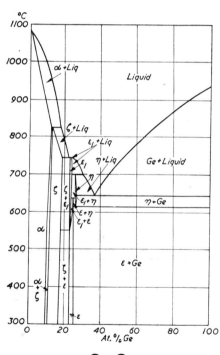

Cu–Ge

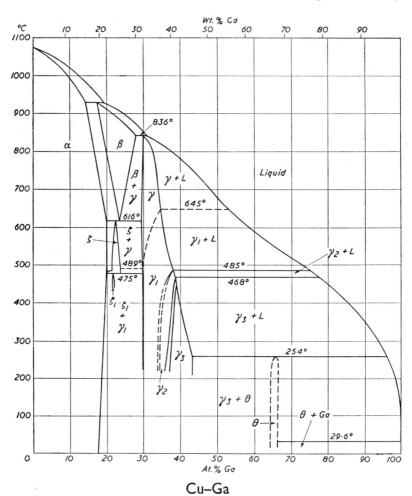

Cu–Ga

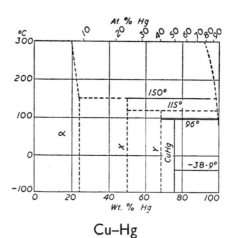

Cu–Hg

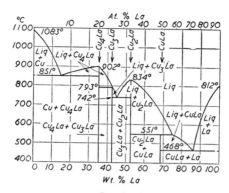

Cu–La

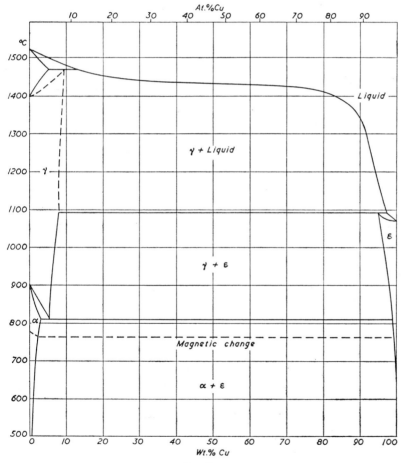

Cu–Fe

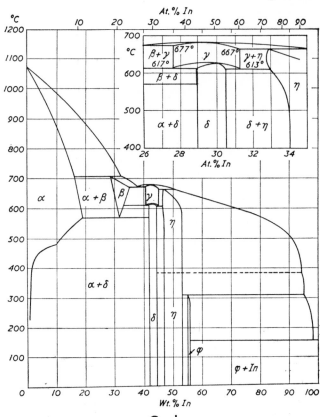

Cu–In

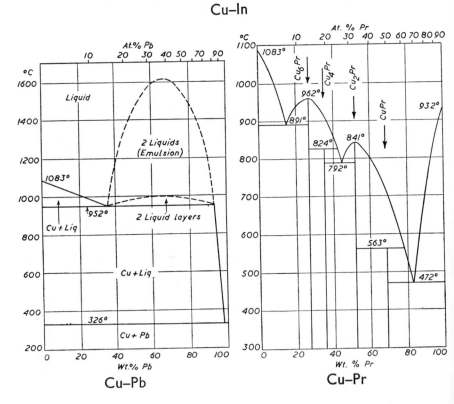

Cu–Pb

Cu–Pr

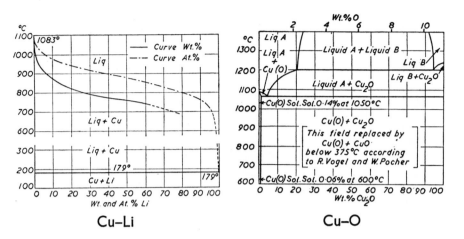

Cu–Li

Cu–O

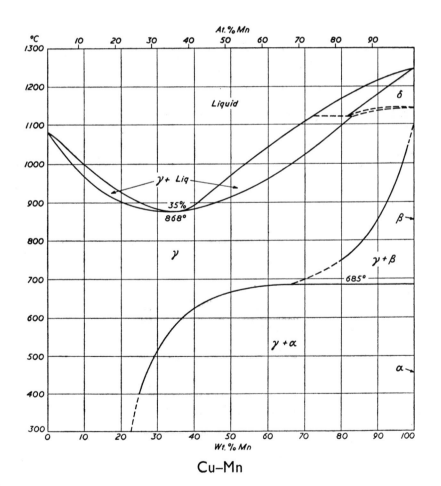

Cu–Mn

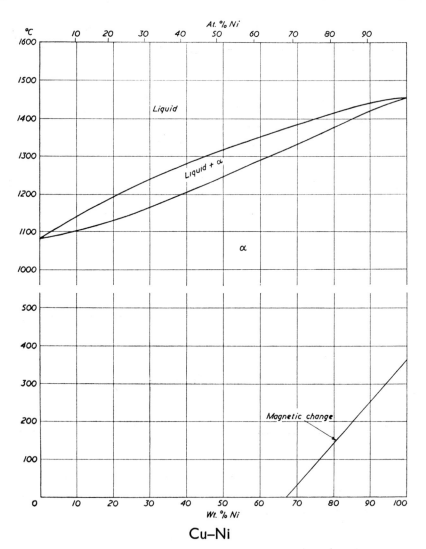

Cu–Ni

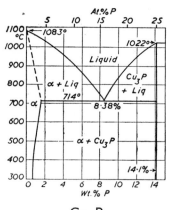

Cu–P

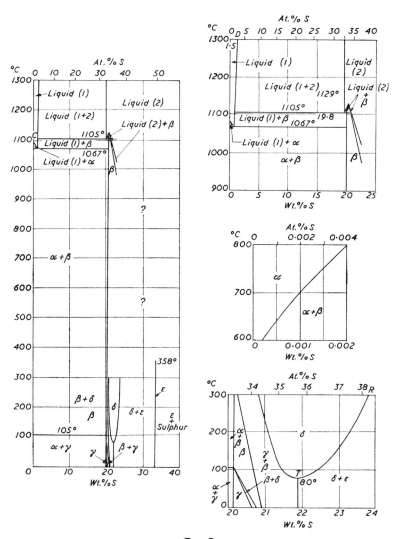

Cu–S

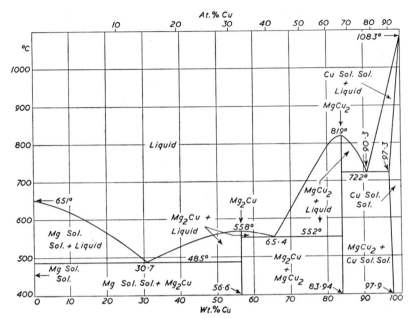

Cu–Mg

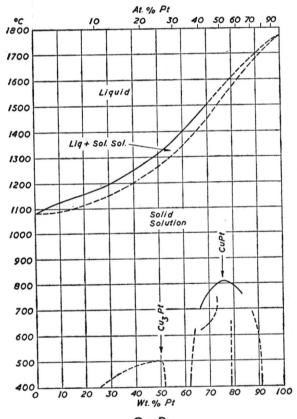

Cu–Pt

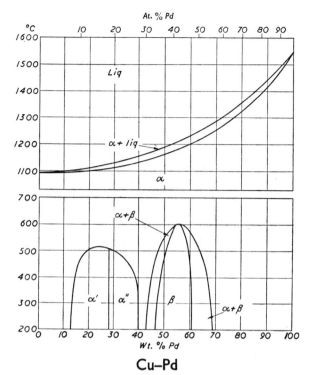

Cu–Pd

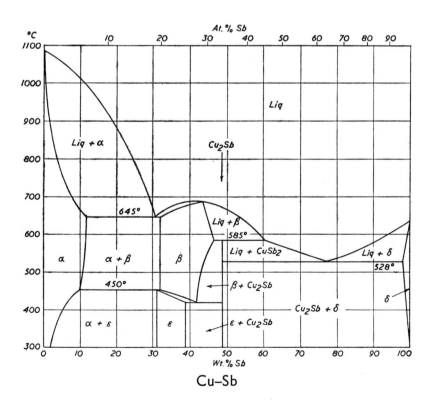

Cu–Sb

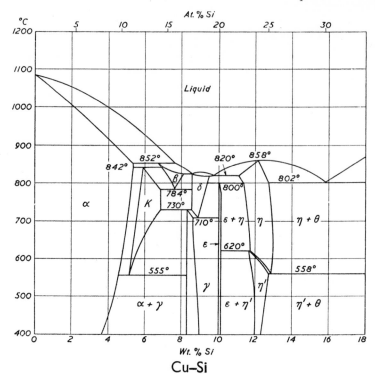

Cu–Si

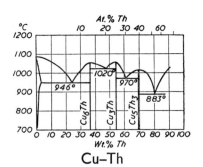

Cu–Th

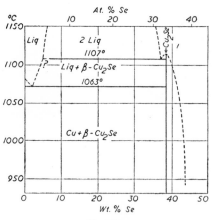

Cu–Se

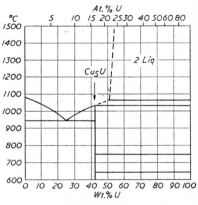

Cu–U

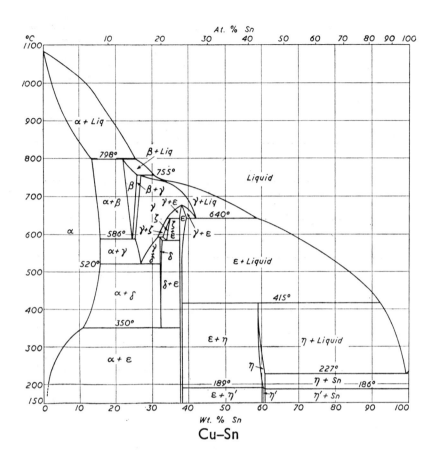

Cu–Sn

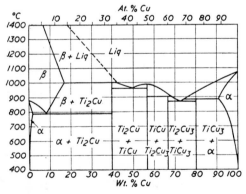

Cu–Ti

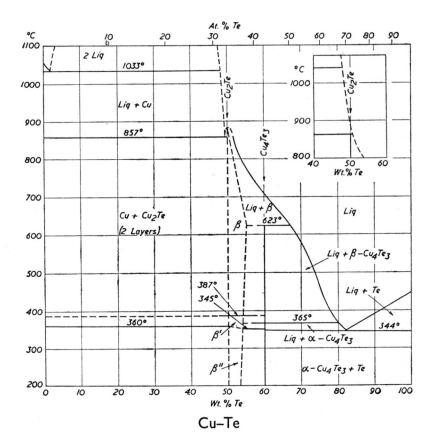

Cu–Te

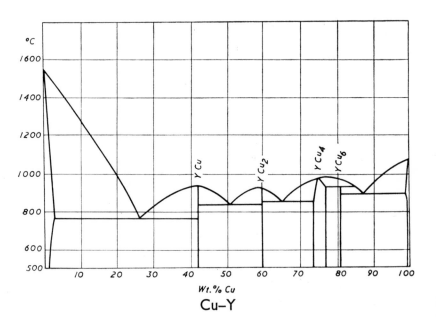

Cu–Y

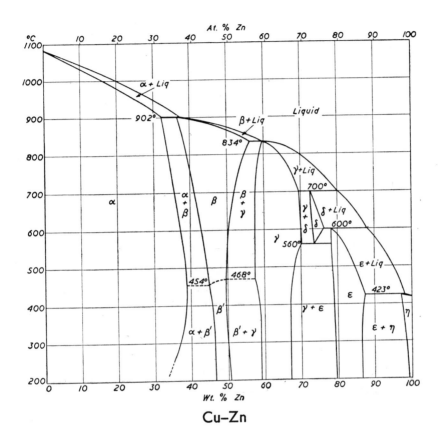

Cu–Zn

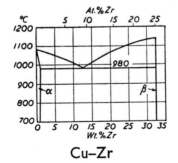

Cu–Zr

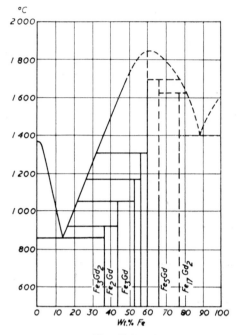

Fe–Gd

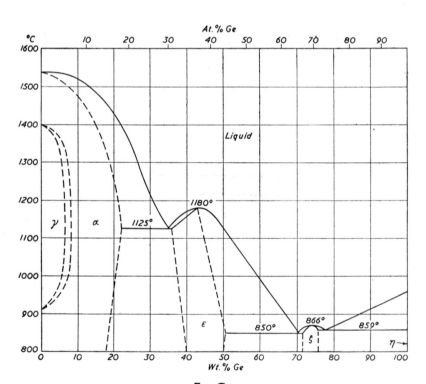

Fe–Ge

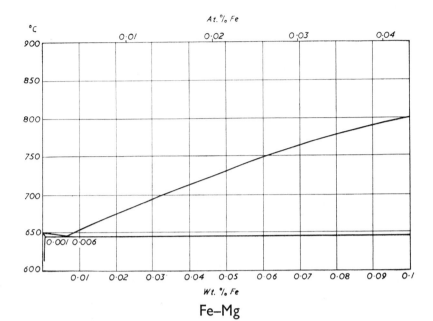

Fe–Mg

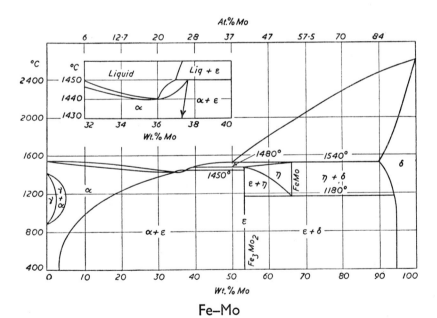

Fe–Mo

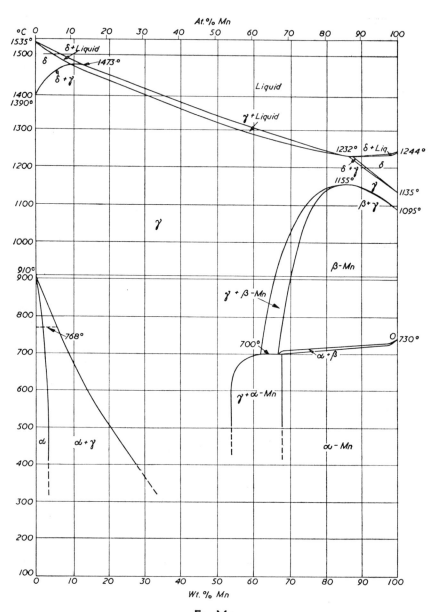

Fe–Mn

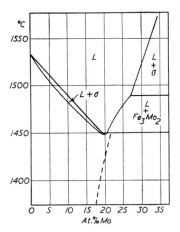

Fe–Mo

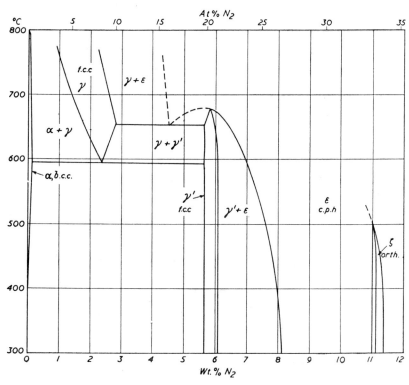

Fe–N

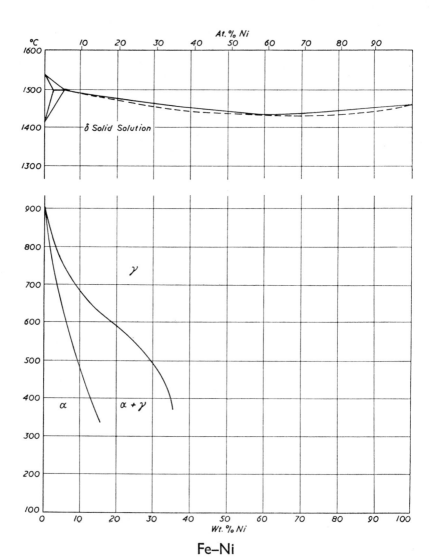

Fe–Ni

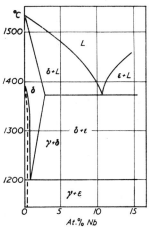

Fe–Nb

Fe–Nb

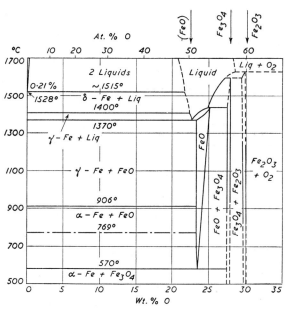

Fe–O

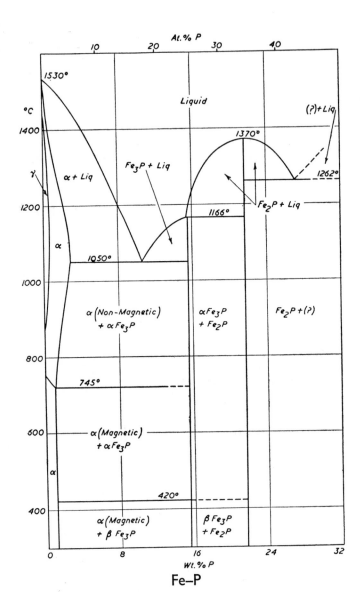

Fe–P

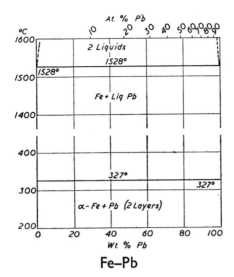

Fe–Pb

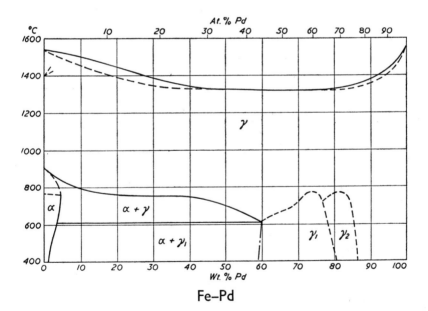

Fe–Pd

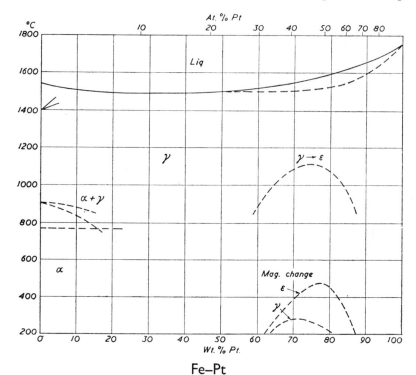

Fe–Pt

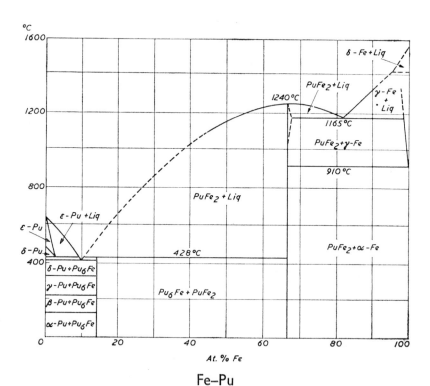

Fe–Pu

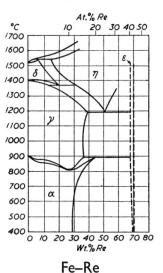

Fe–Re

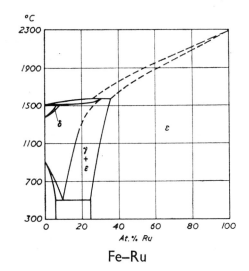

Fe–Ru

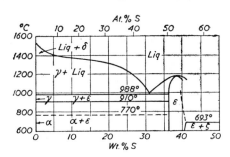

Fe–S

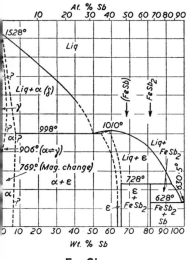

Fe–Sb

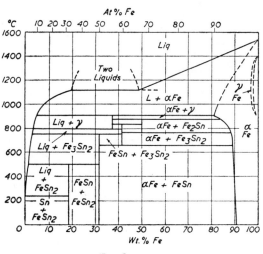

Fe–Sn

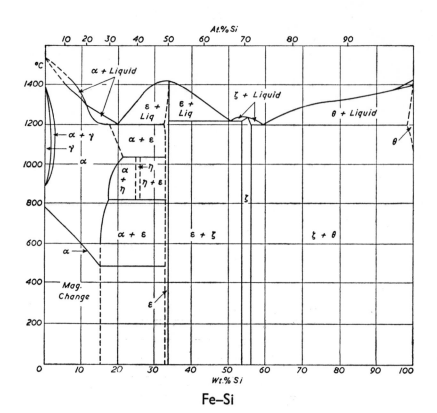

Fe–Si

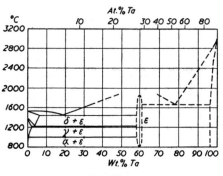

Fe–Ta

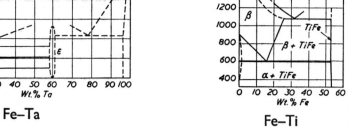

Fe–Ti

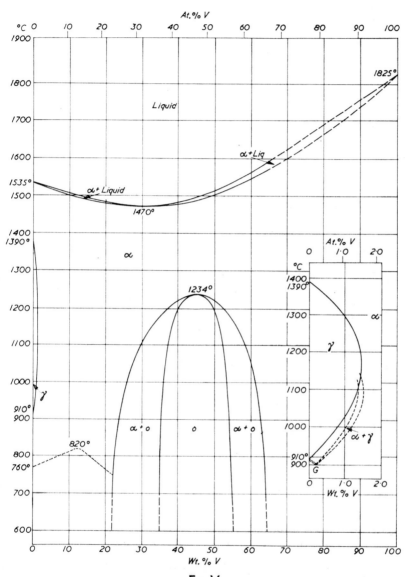

Fe–V

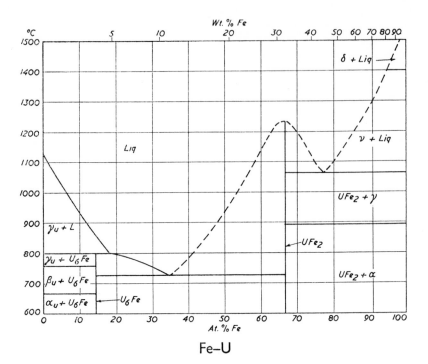

Fe–U

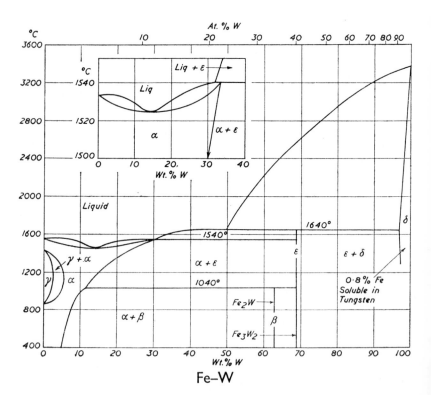

Fe–W

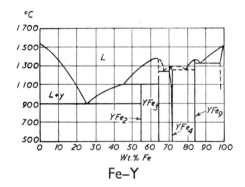

Fe–Y

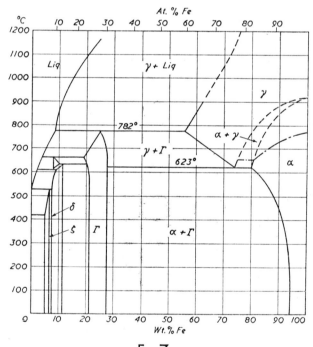

Fe–Zn

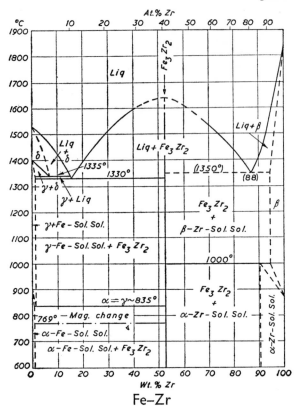

Fe–Zr

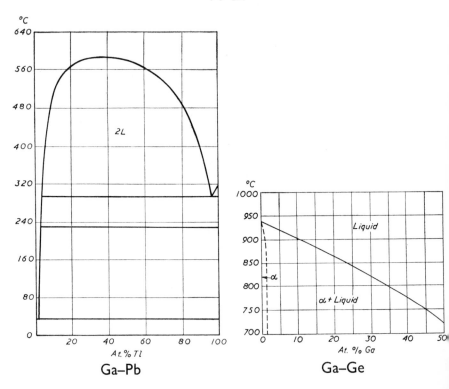

Ga–Pb

Ga–Ge

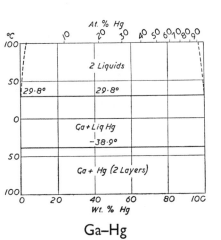

Ga–Hg

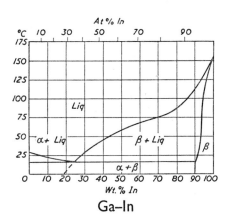

Ga–In

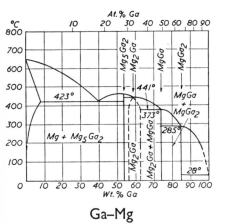

Ga–Mg

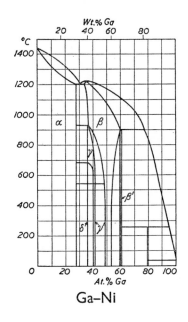

Ga–Ni

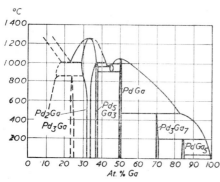

Ga–Pd

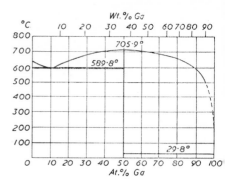

Ga–Sb

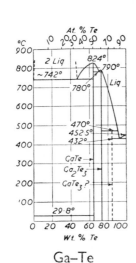

Ga–Te

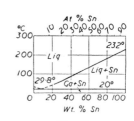

Ga–Sn

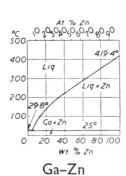

Ga–Zn

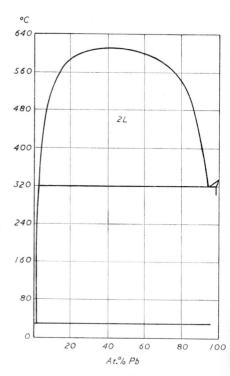

Ga–Tl

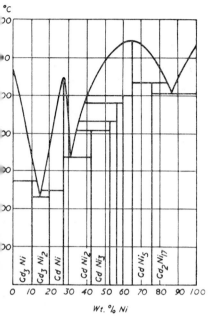

Gd–Ni

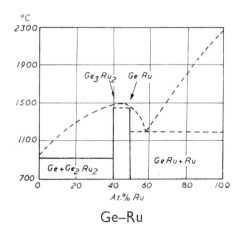

Ge–Mn

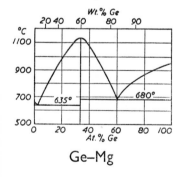

Ge–Mg

Ge–Ru

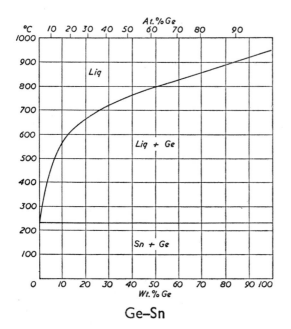

Ge–Sn

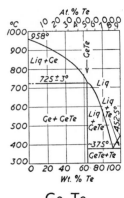

Ge–Te

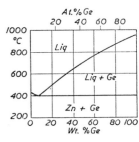

Ge–Zn

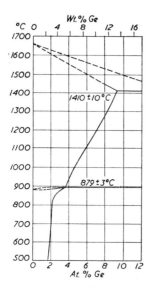

Ge–Ti

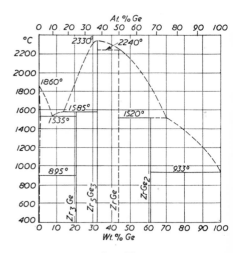

Ge–Zr

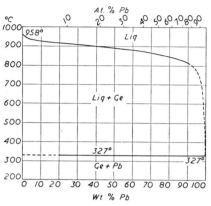

Ge–Pb

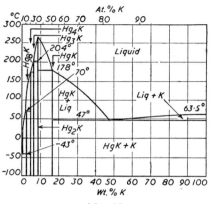

Hg–K

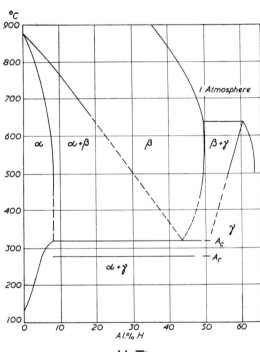

H–Ti

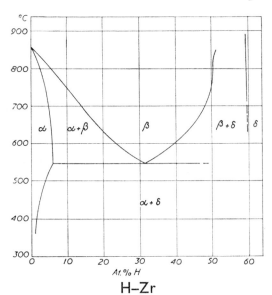

H–Zr

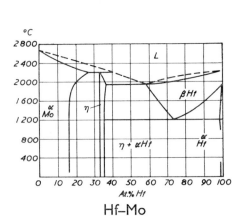

Hf–Mo

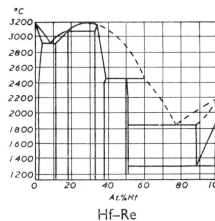

Hf–Re

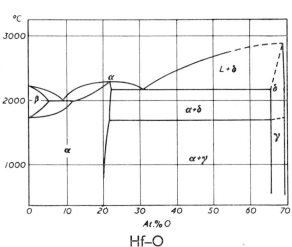

Hf–O

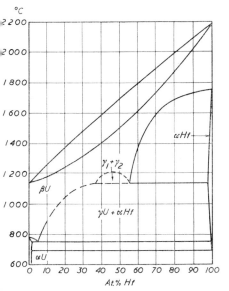

Hf–U

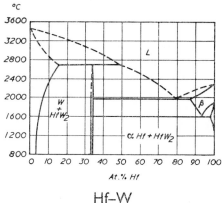

Hf–W

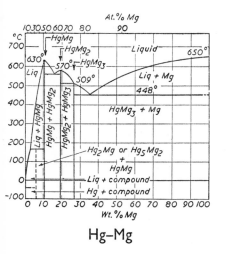

Hg–Mg

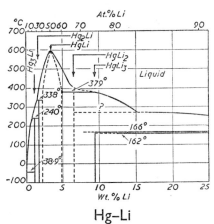

Hg–Li

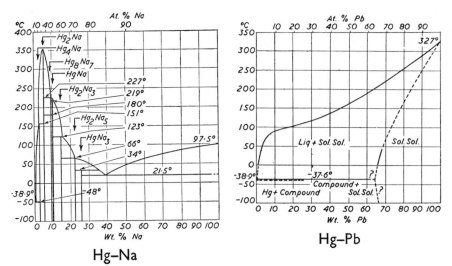

Hg–Na

Hg–Pb

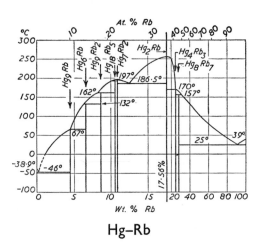

Hg–Rb

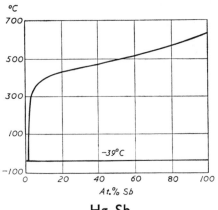

Hg–Sb

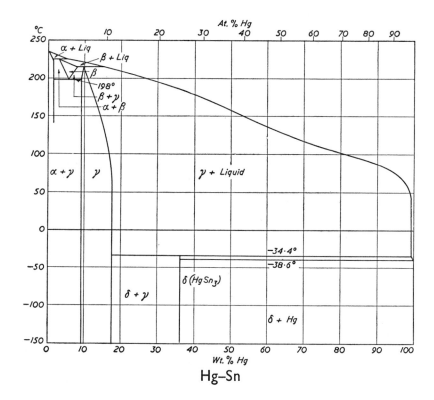

Hg–Sn

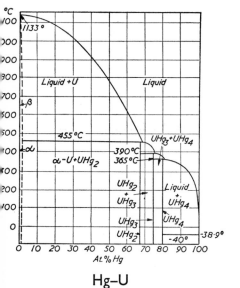

Hg–U

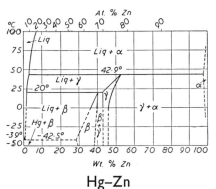

Hg–Zn

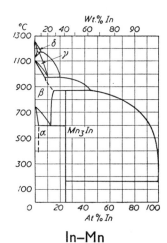

In–Mn

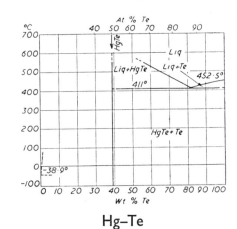

Hg–Te

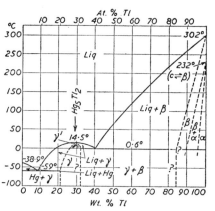

Hg–Tl

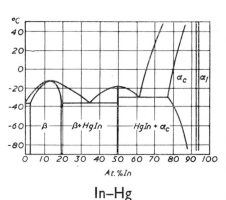

In–Hg

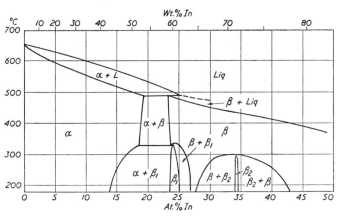

In–Mg

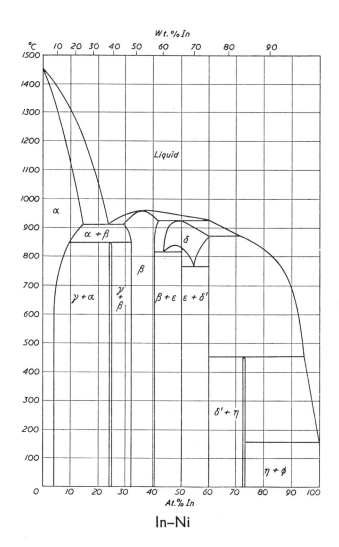

In–Ni

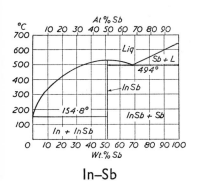

In–Sb

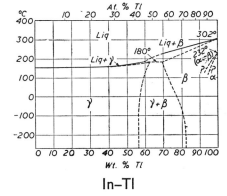

In–Tl

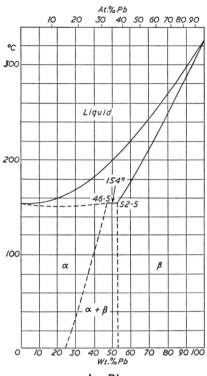

In–Pb

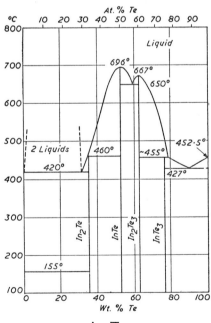

In–Te

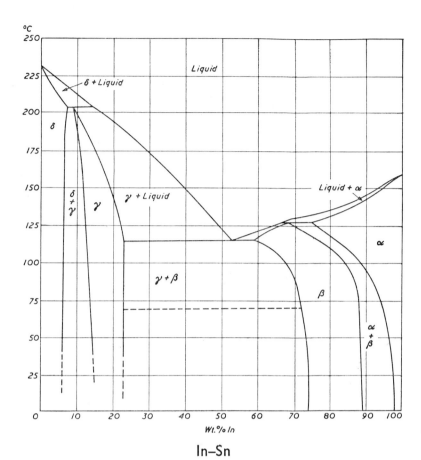

In–Sn

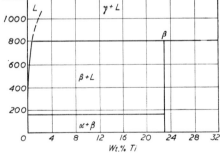

In–Ti

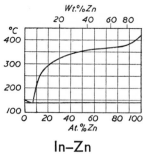

In–Zn

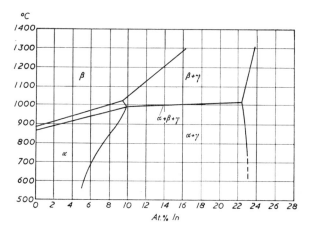

In–Zr

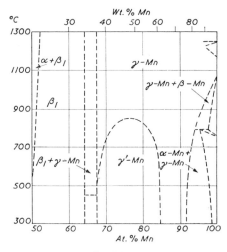

Ir–Mn

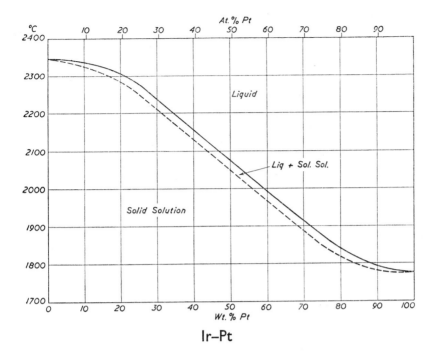

Ir–Pt

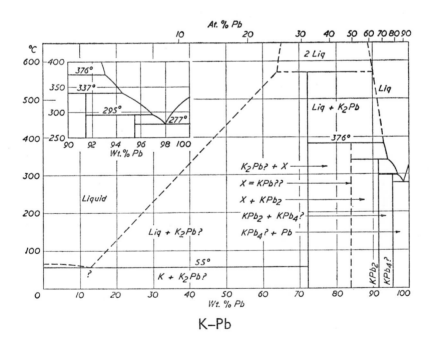

K–Pb

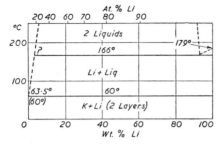

K–Li

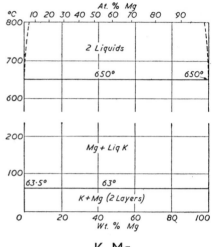

K–Mg

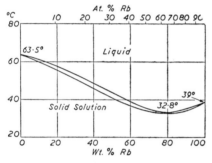

K–Rb

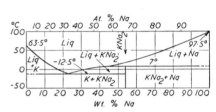

K–Na

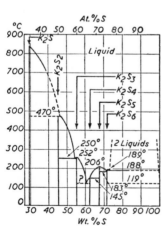

K–S

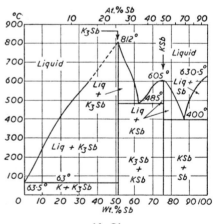

K–Sb

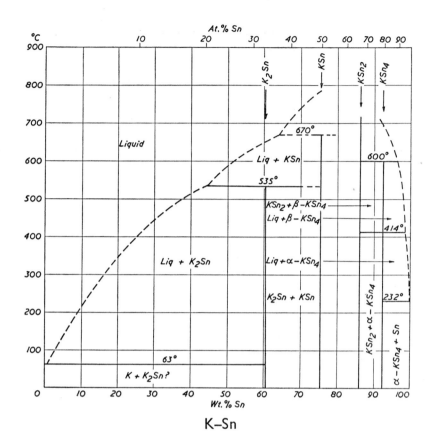

K–Sn

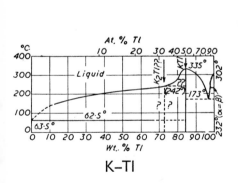

K–Tl

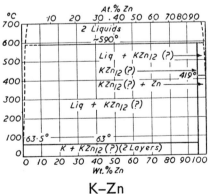

K–Zn

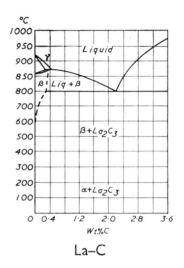

La–C

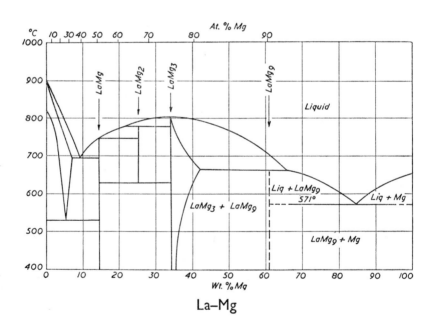

La–Mg

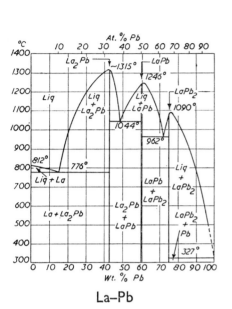

La–Pb

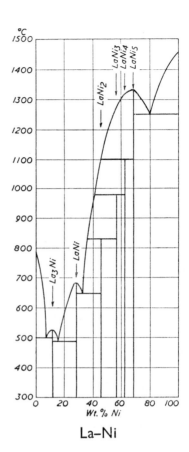

La–Ni

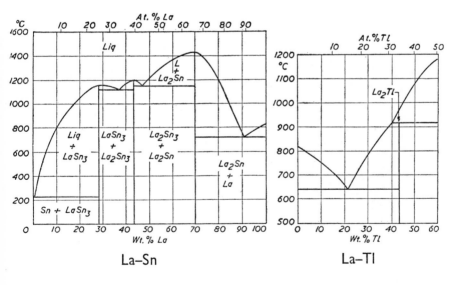

La–Sn

La–Tl

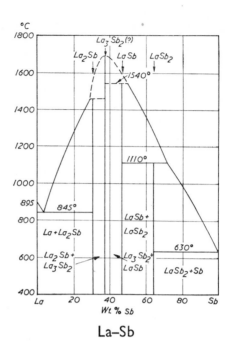

La–Sb

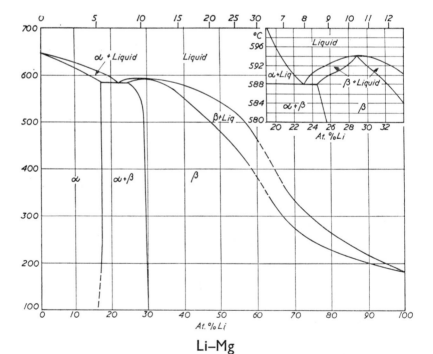

Li–Mg

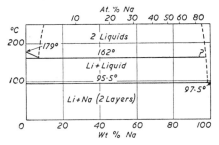

Li–Na

Li–S

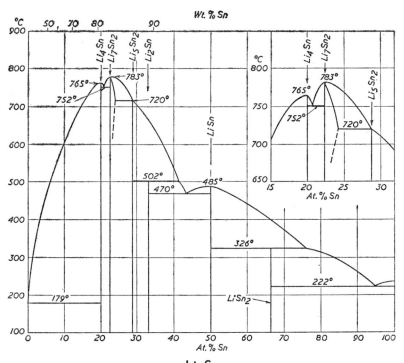

Li–Pb

Li–Sn

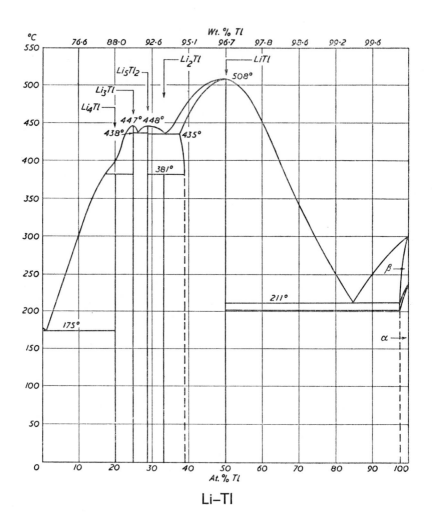

Li–Tl

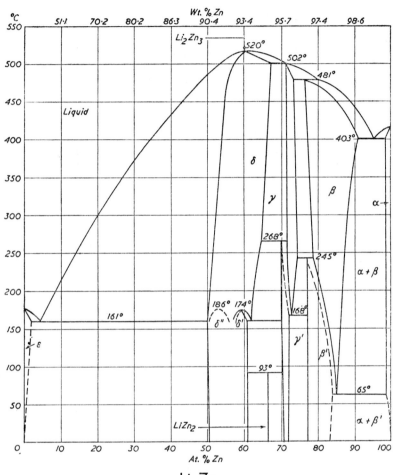

Li–Zn

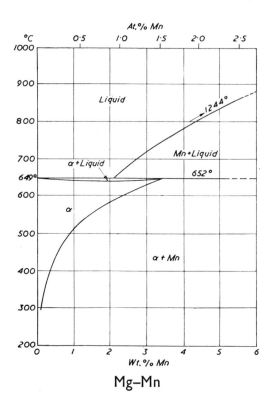

Mg–Mn

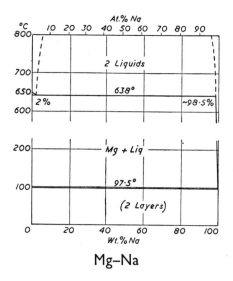

Mg–Na

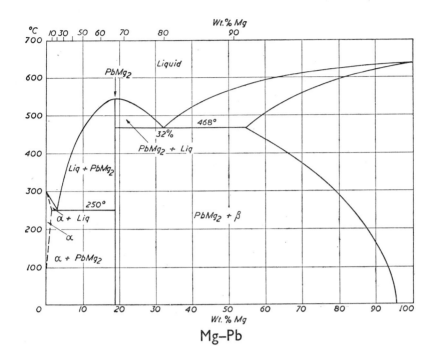

Mg–Pb

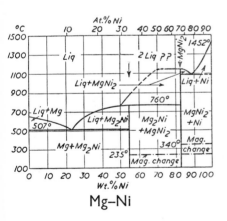

Mg–Ni

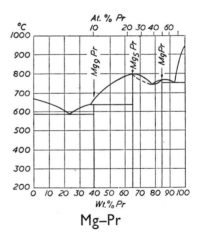

Mg–Pr

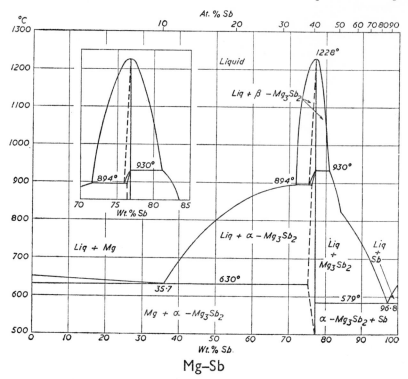

Mg–Sb

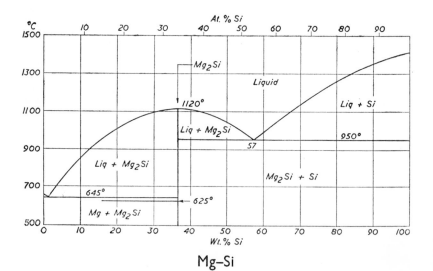

Mg–Si

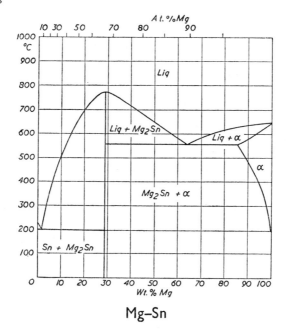

Mg–Sn

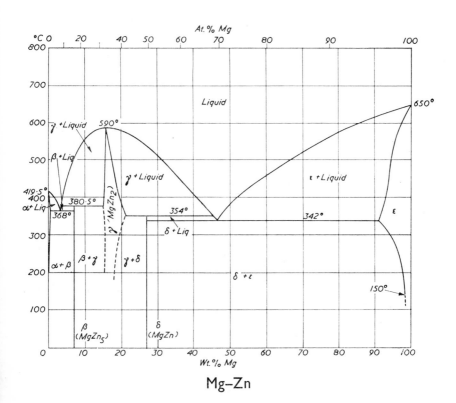

Mg–Zn

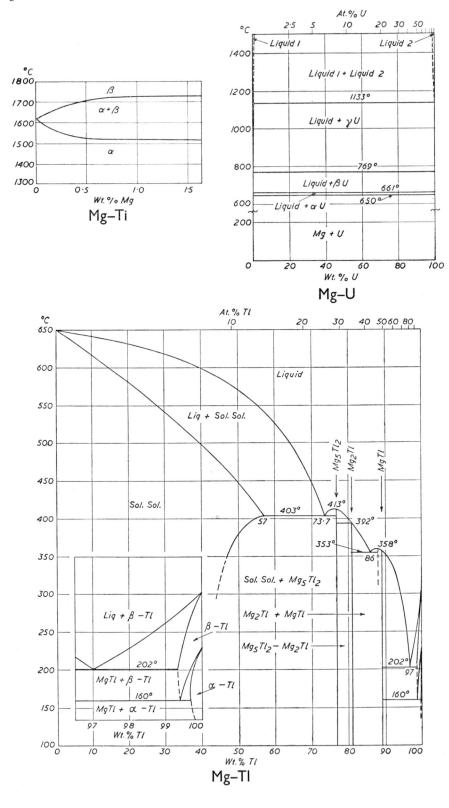

Mg–Ti

Mg–U

Mg–Tl

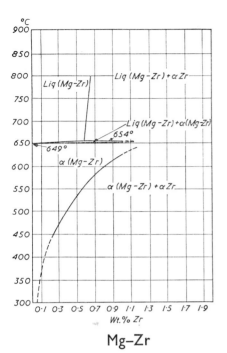

Mg–Zr

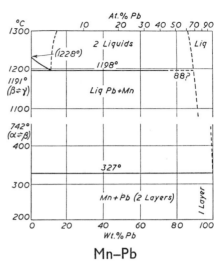

Mn–Pb

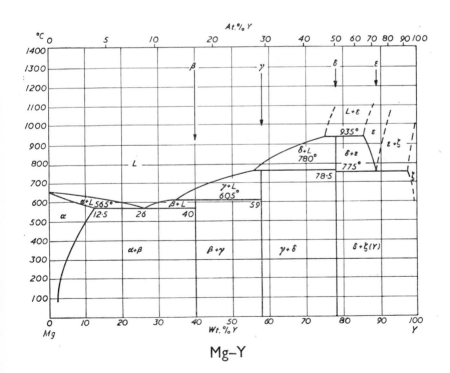

Mg–Y

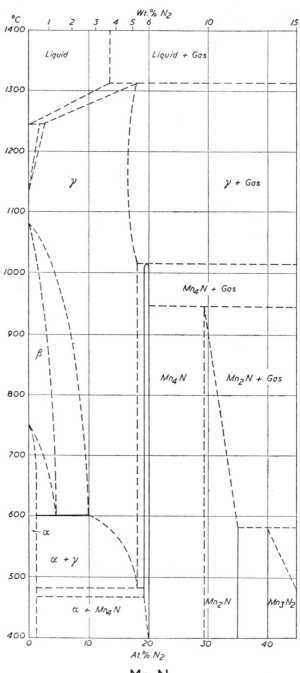

Mn–N

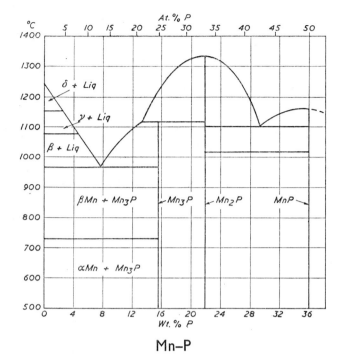

Mn–P

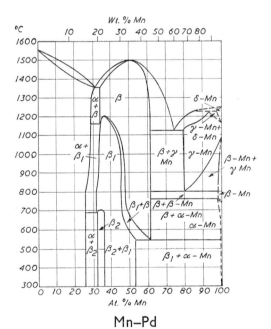

Mn–Pd

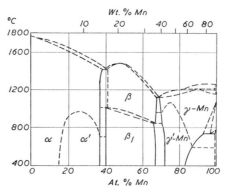

Mn–Pt

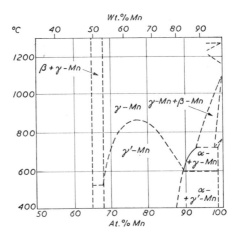

Mn–Rh

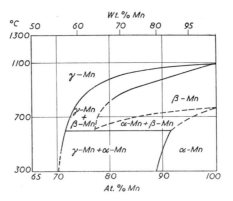

Mn–Ru

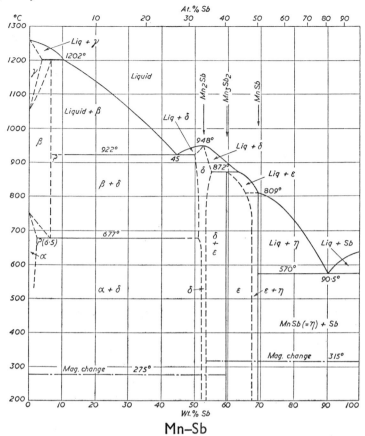

Mn–Sb

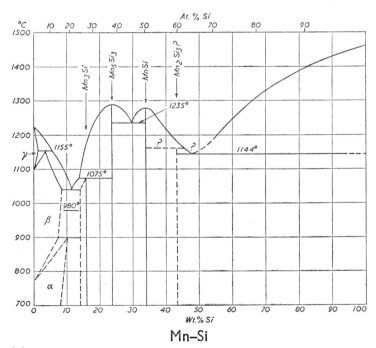

Mn–Si

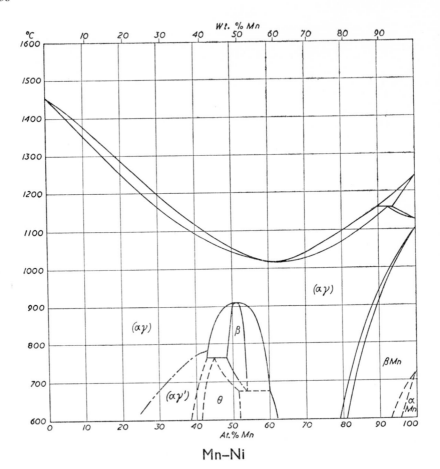

Mn–Ni

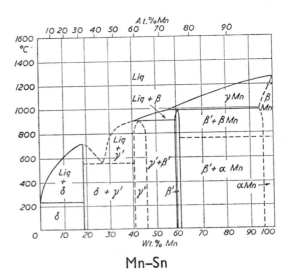

Mn–Sn

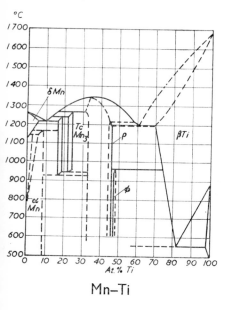

Mn–Ti

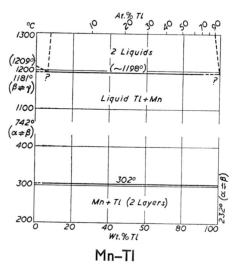

Mn–Tl

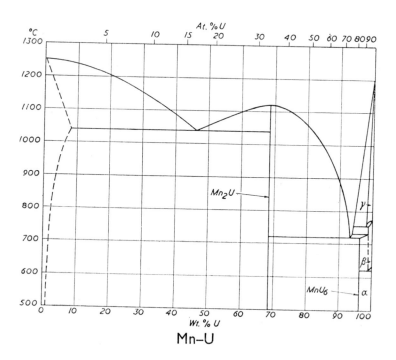

Mn–U

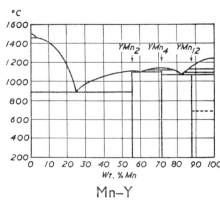

Mn–Y

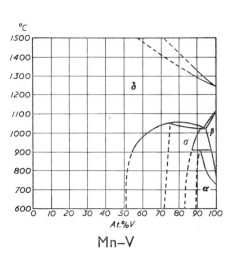

Mn–V

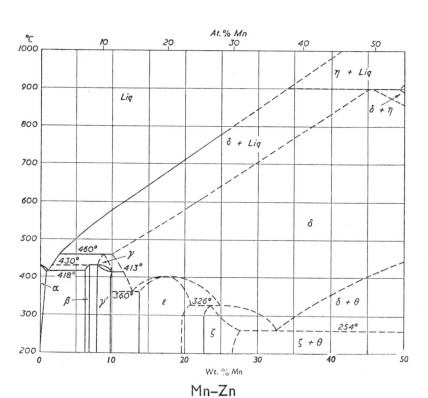

Mn–Zn

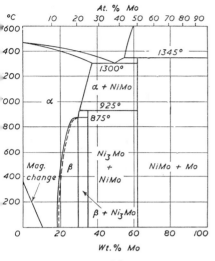

Mo–Ni

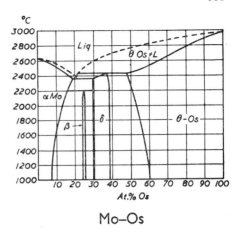

Mo–Os

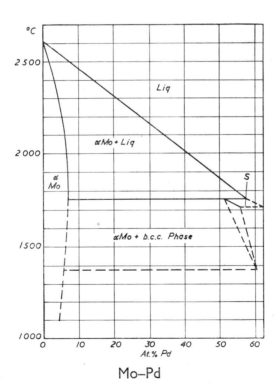

Mo–Pd

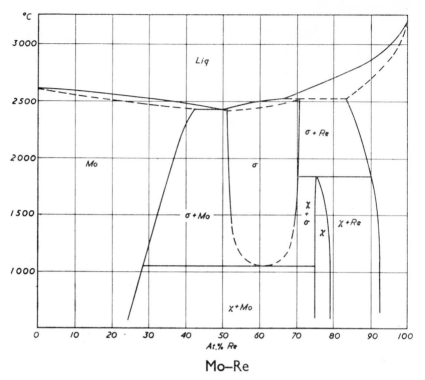

Mo–Re

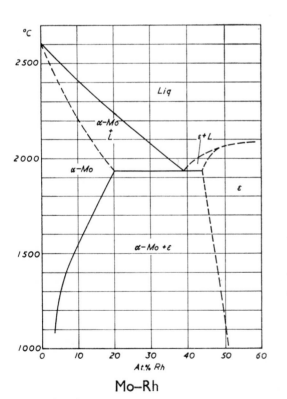

Mo–Rh

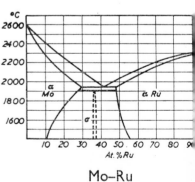

Mo–Ru

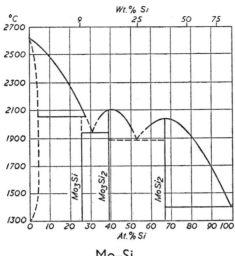

Mo–Si

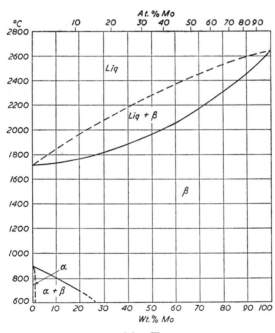

Mo–Ti

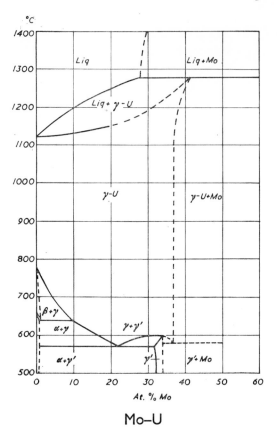

Mo–U

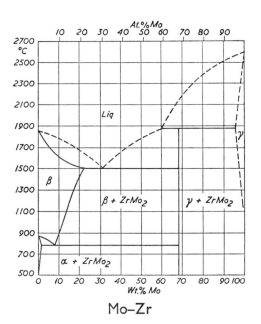

Mo–Zr

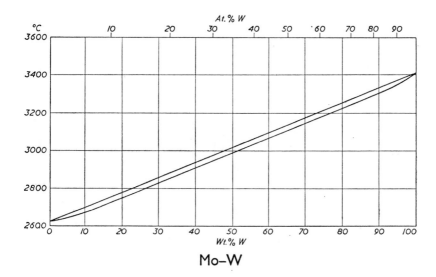

Mo–W

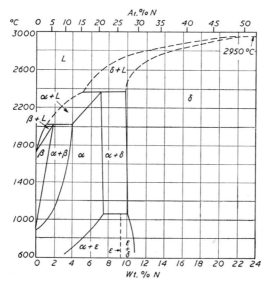

N–Ti

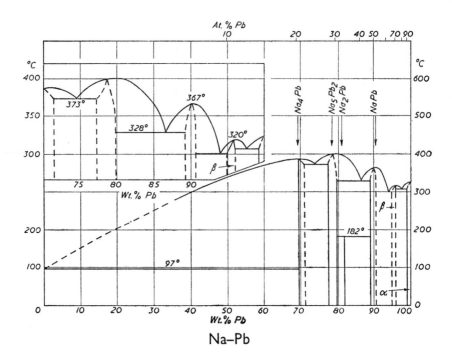

Na–Pb

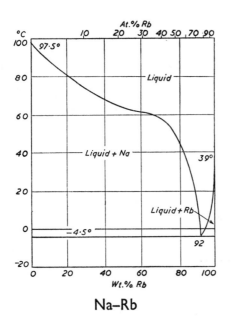

Na–Rb

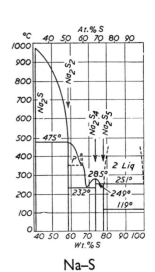

Na–S

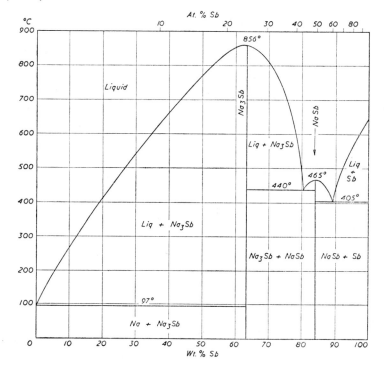

Na–Sb

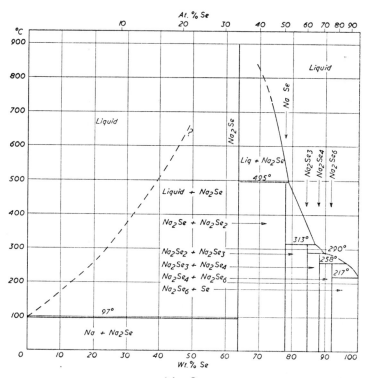

Na–Se

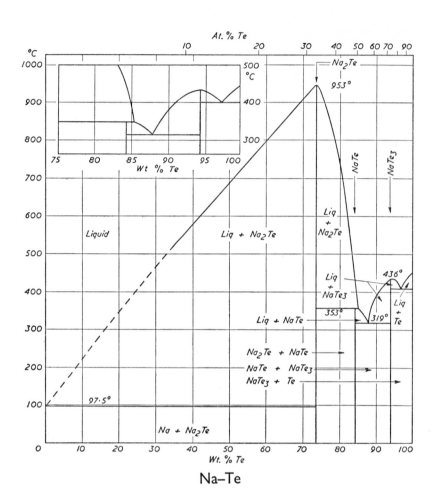

Na–Te

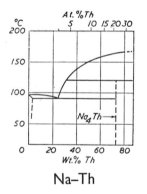

Na–Th

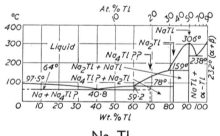

Na–Tl

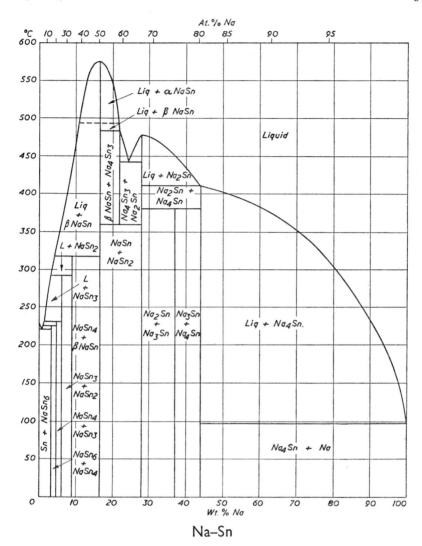

Na–Sn

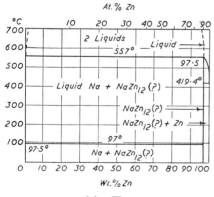

Na–Zn

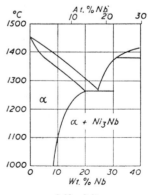

Nb–Ni

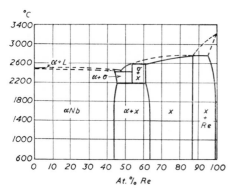

Nb–Re

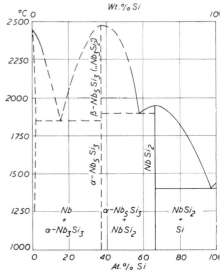

Nb–Si

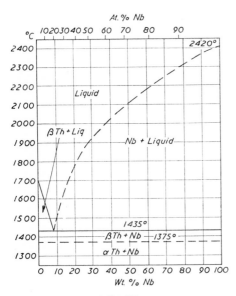

Nb–Th

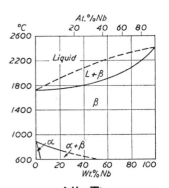

Nb–Ti

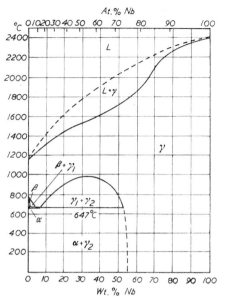

Nb–U

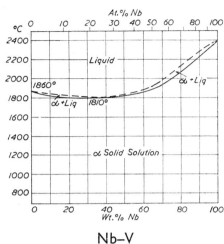

Nb–V

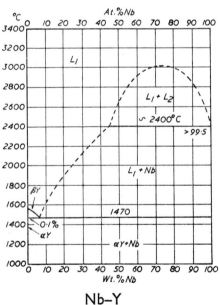

Nb–Y

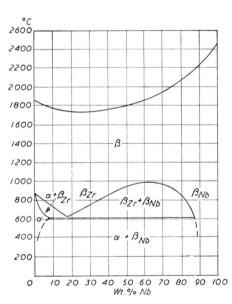

Nb–Zr

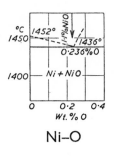

Ni–O

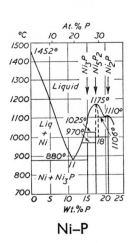

Ni–P

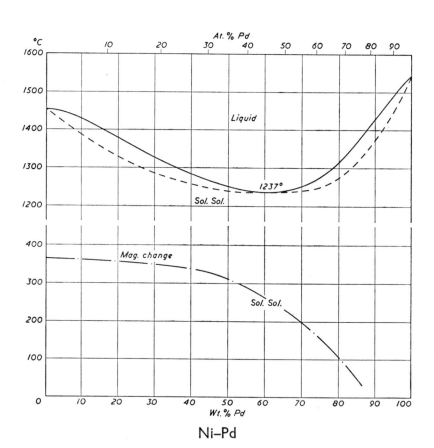

Ni–Pd

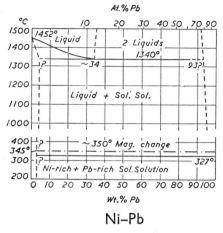

Ni–Pb

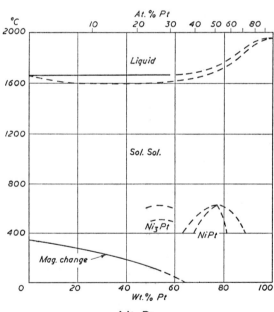

Ni–Pt

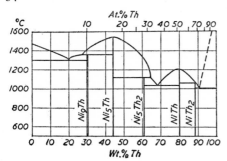

Ni–Th

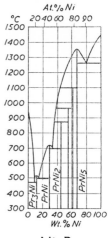

Ni–Pr

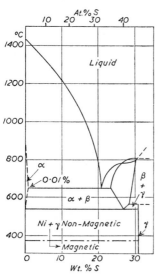

Ni–S

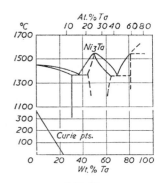

Ni–Ta

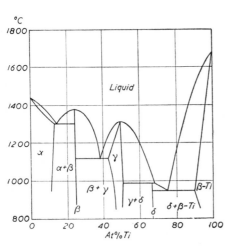

Ni–Ti

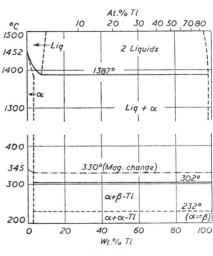

Ni–Tl

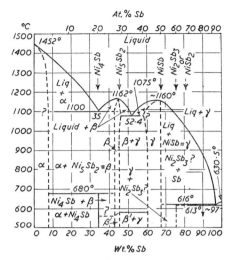

Ni–Sb

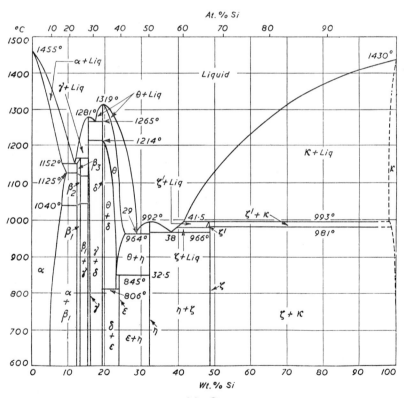

Ni–Si

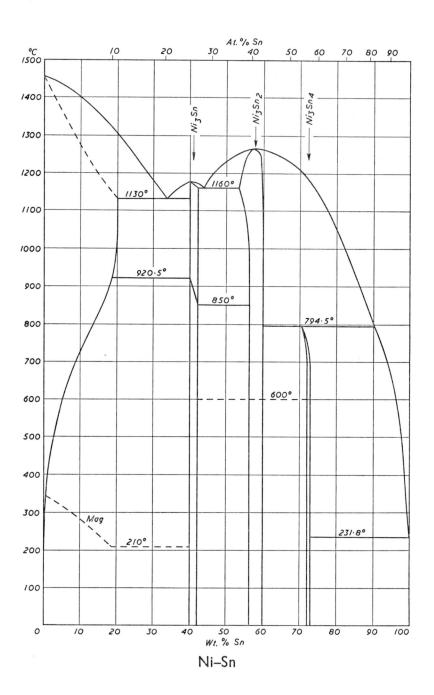

Ni–Sn

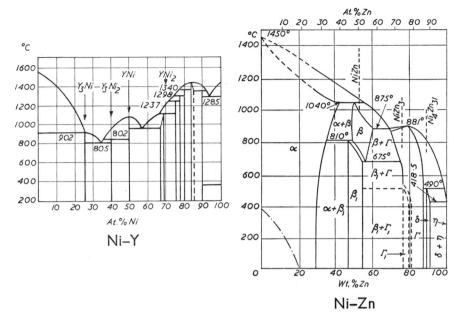

Ni–Y

Ni–Zn

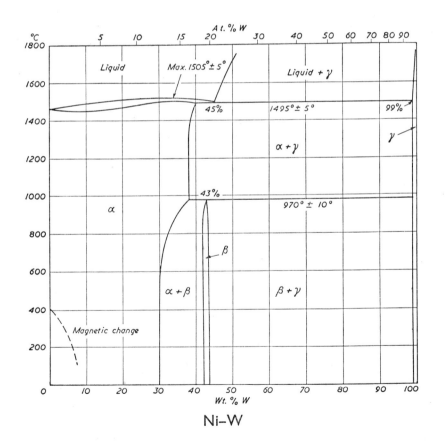

Ni–W

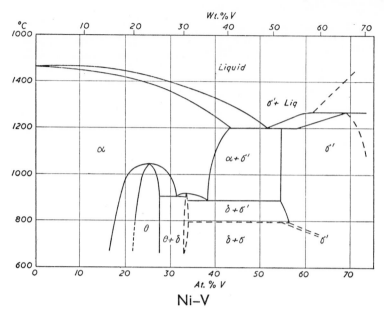

Ni–V

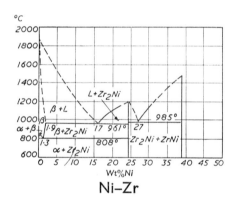

Ni–Zr

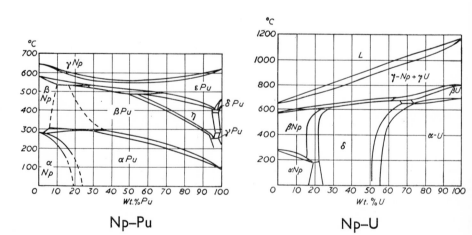

Np–Pu   Np–U

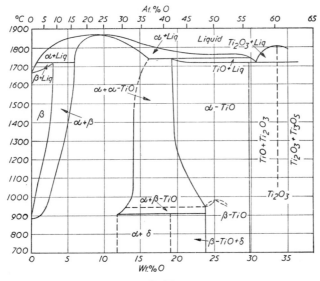

O–Ti

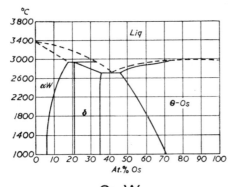

Os–W

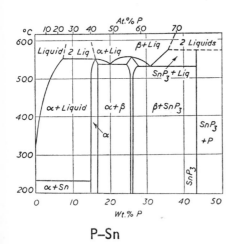

P–Sn

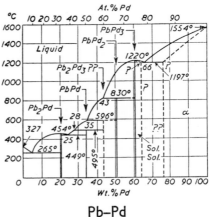

Pb–Pd

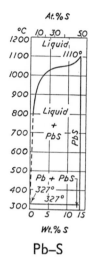

Pb–S

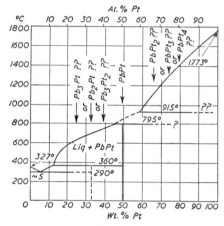

Pb–Pt

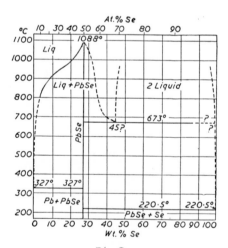

Pb–Se

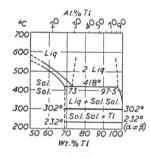

P–Tl

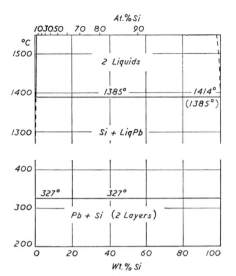

Pb–Si

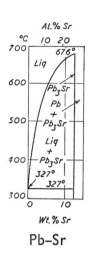

Pb–Sr

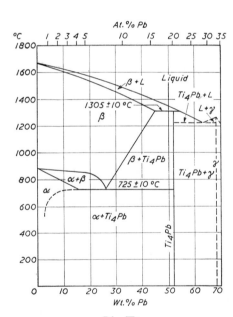

Pb–Ti

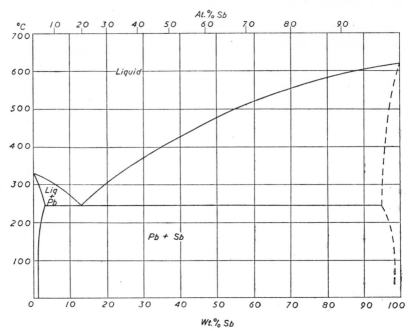

Pb–Sb

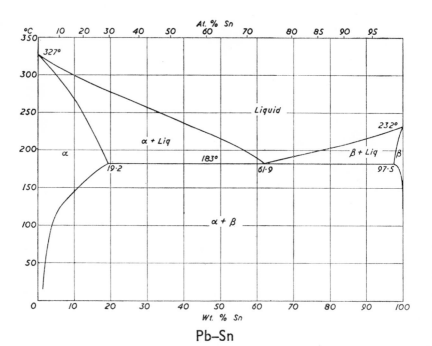

Pb–Sn

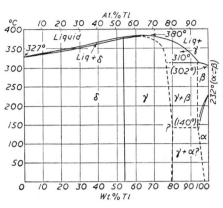

Pb–Tl

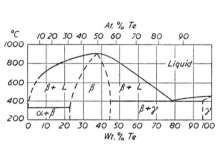

Pb–Te

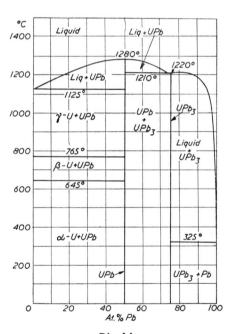

Pb–U

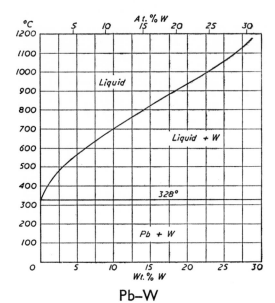

Pb–W

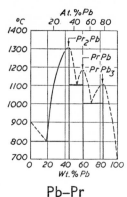

Pb–Pr

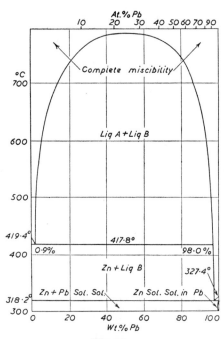

Pb–Zn

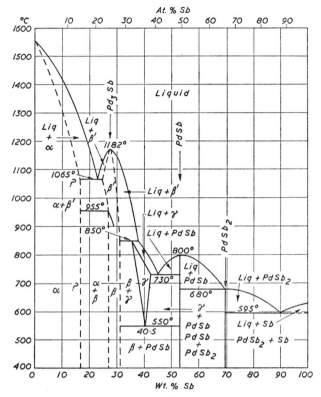

Pd–Sb

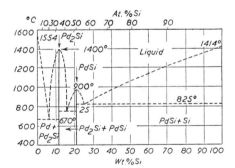

Pd–Si

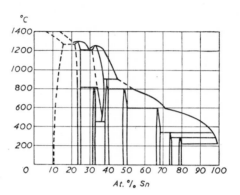

Pd–Sn

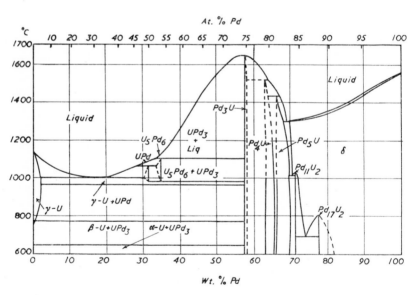

Pd–U

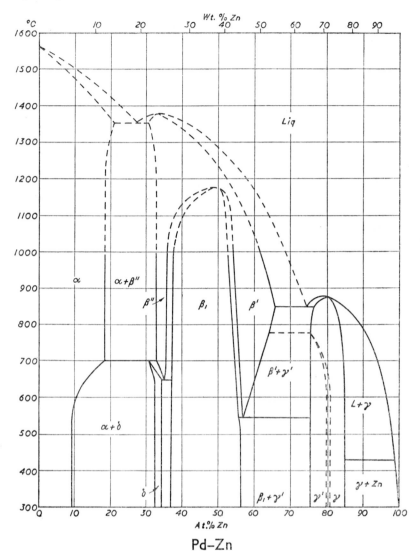

Pd–Zn

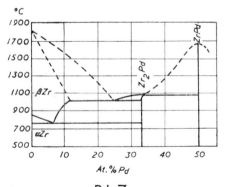

Pd–Zr

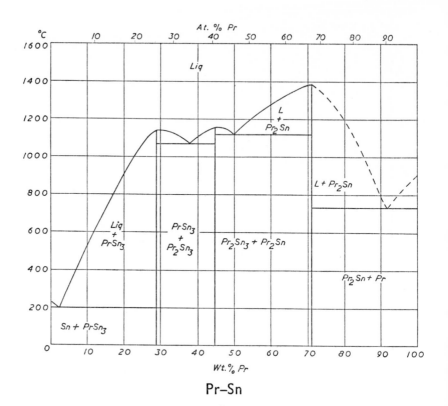

Pr–Sn

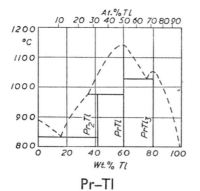

Pr–Tl

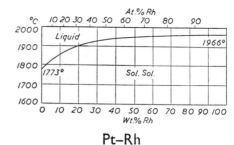

Pt–Rh

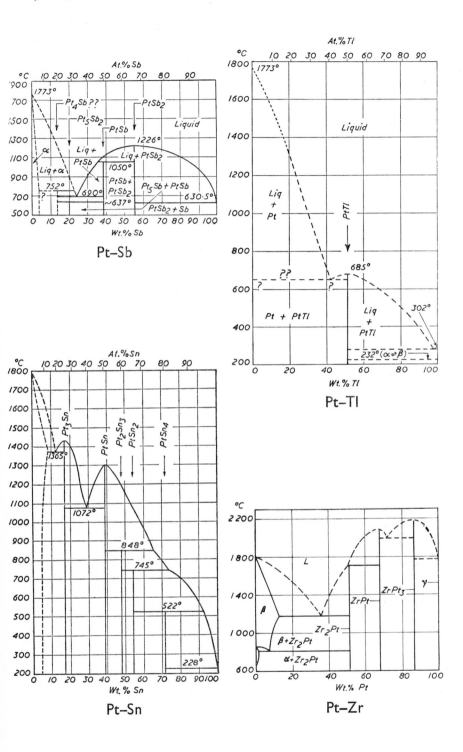

Pt–Sb

Pt–Tl

Pt–Sn

Pt–Zr

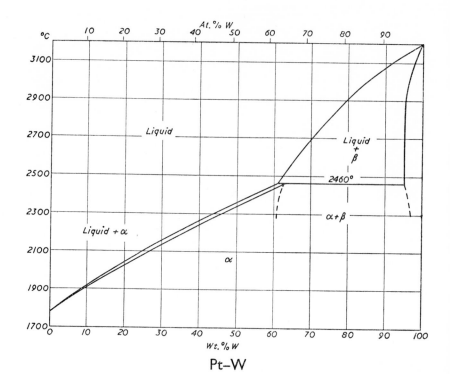

Pt–W

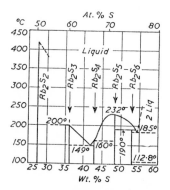

Rb–S

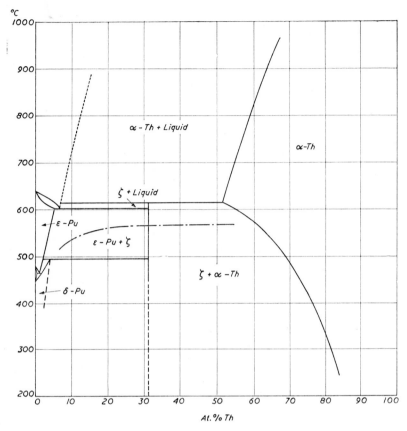

Pu–Th

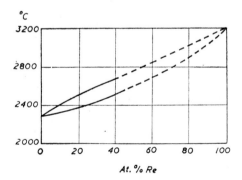

Re–Ru

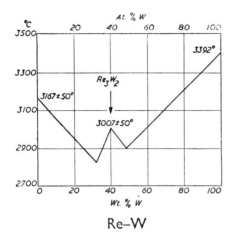

Re–W

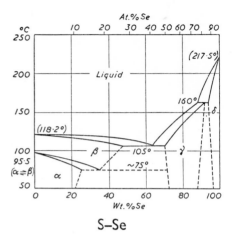

S–Se

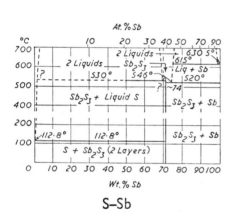

S–Sb

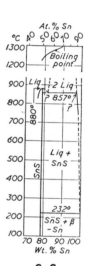

S–Sn

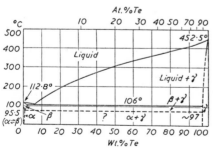

S–Te

S–Tl

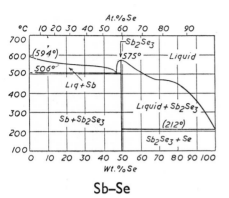

Sb–Se

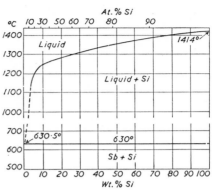

Sb–Si

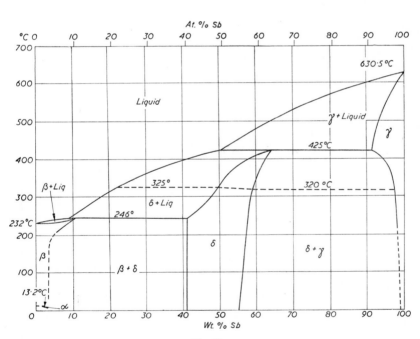

Sb–Sn

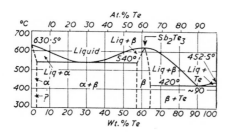

Sb–Te

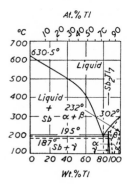

Sb–Tl

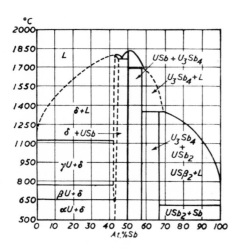

Sb–U

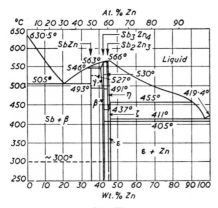

Sb–Zn

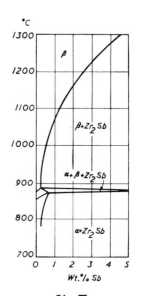

Sb–Zr

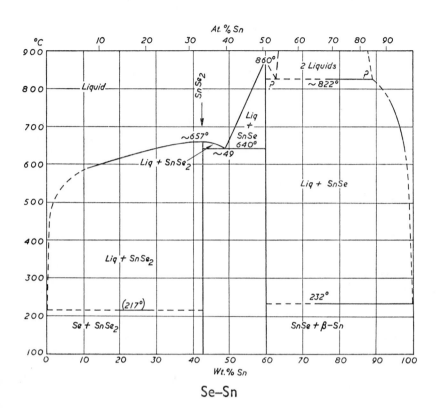

Se–Sn

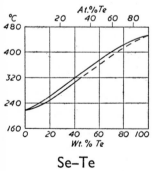

Se–Te

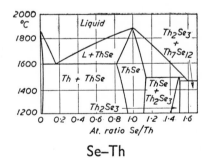

Se–Th

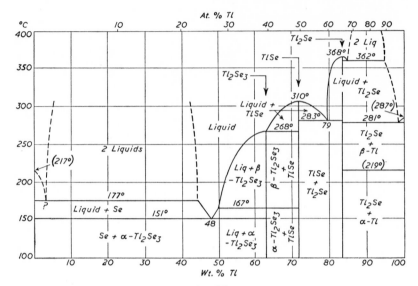

Se–Tl

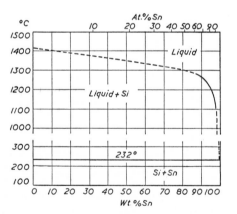

Si–Sn

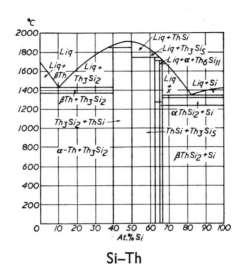

Si–Th

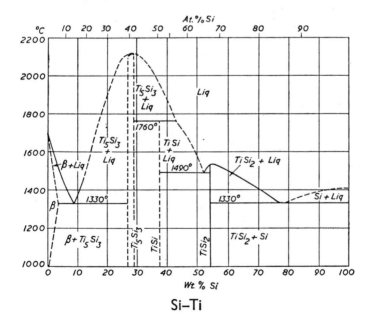

Si–Ti

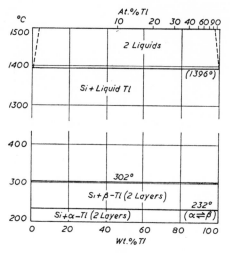

Si–Tl

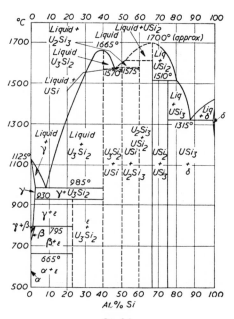

Si–U

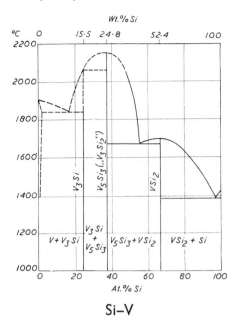

Si–V

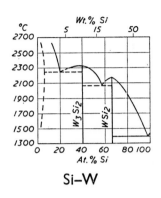

Si–W

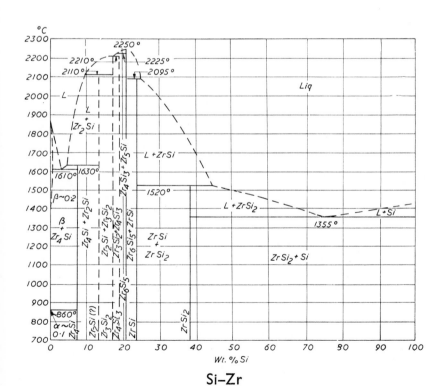

Si–Zr

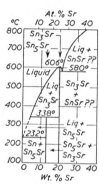

Sn–Sr

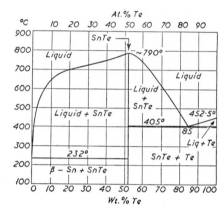

Sn–Te

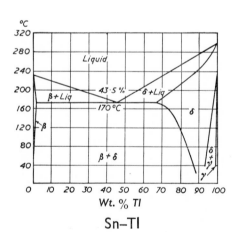

Wt. % Tl

Sn–Tl

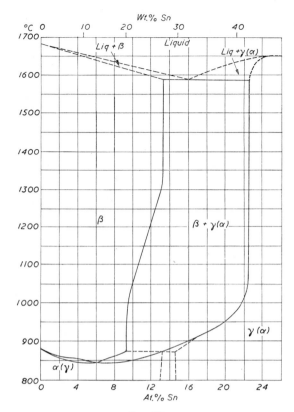

Sn–Ti

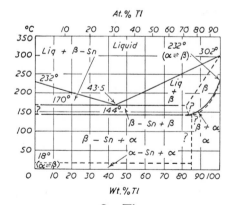

Sn–Tl

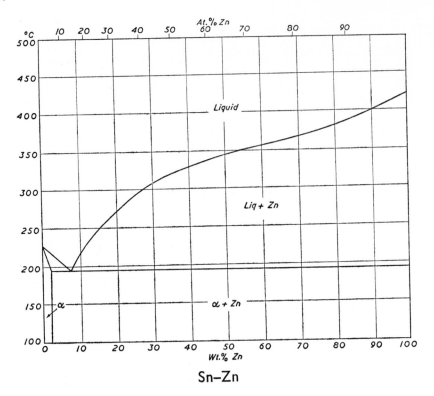

Sn–Zn

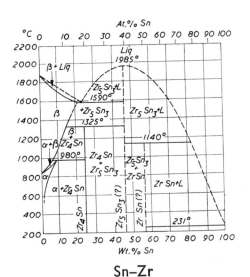

Sn–Zr

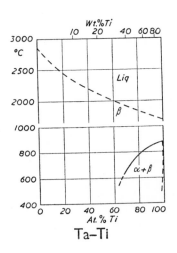

Ta–Ti

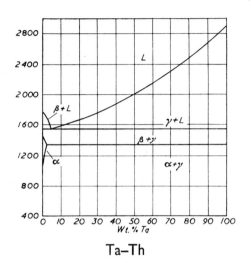

Ta–Th

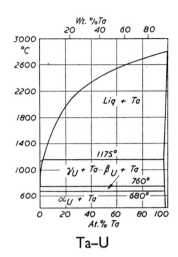

Ta–U

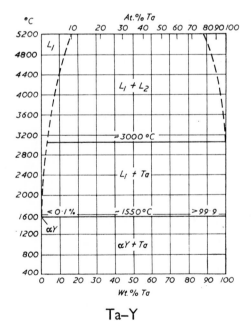

Ta–Y

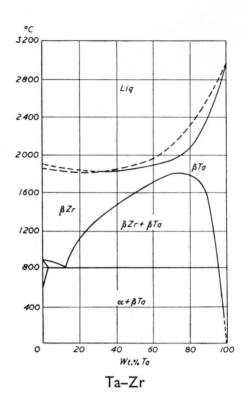

Ta–Zr

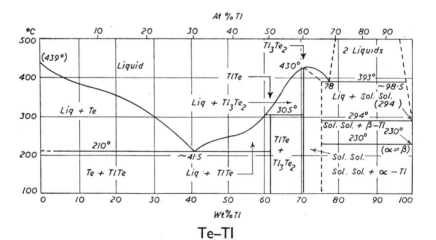

Te–Tl

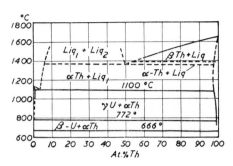

Te–Zn

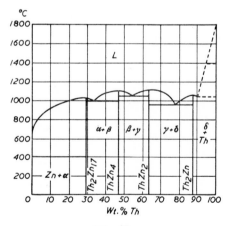

Th–U

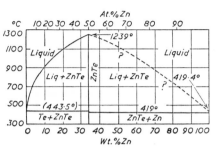

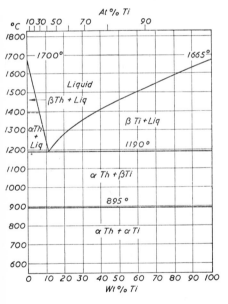

Th–Ti

Th–Zn

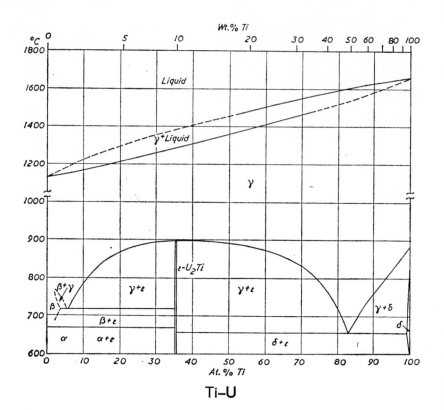

Ti–U

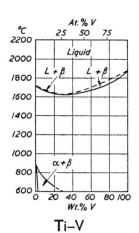

Ti–V

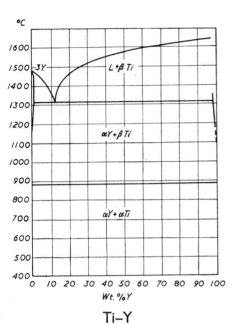

Ti–Y

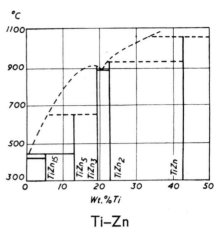

Ti–Zn

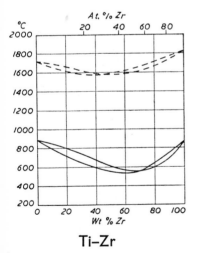

Ti–Zr

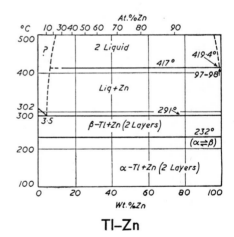

Tl–Zn

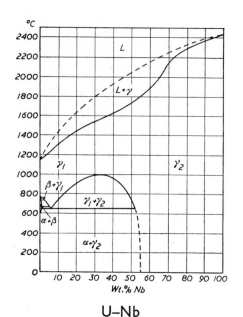

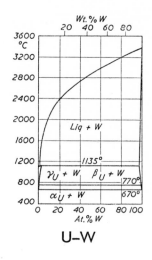

U–Nb

U–W

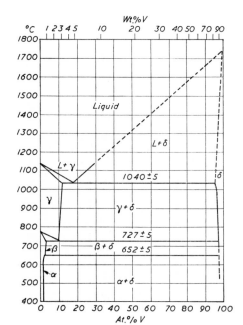

U–V

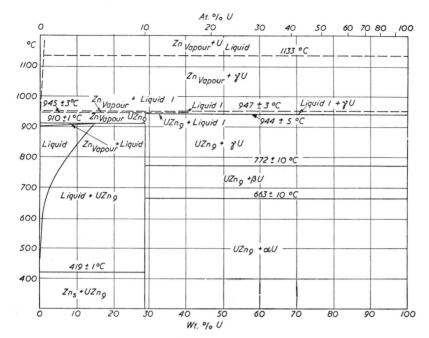

U–Zn at 1 atm

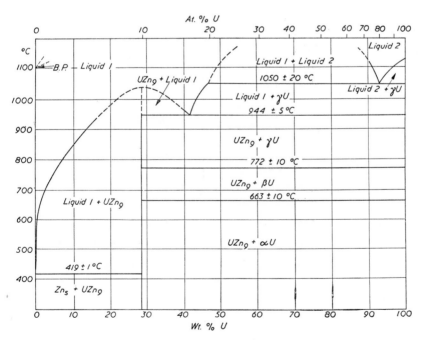

U–Zn at 5 atm

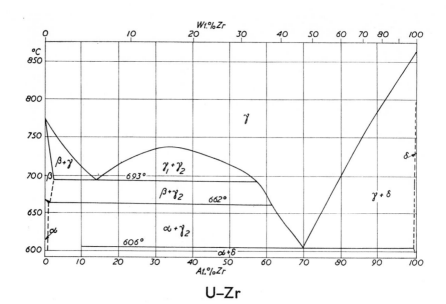

U–Zr

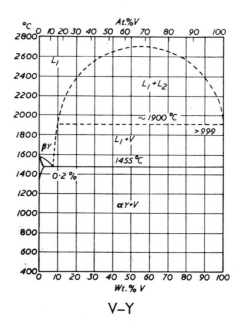

V–Y

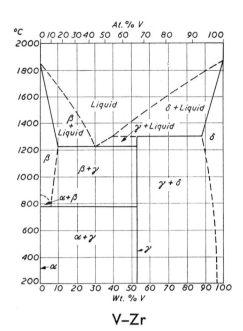

V–Zr

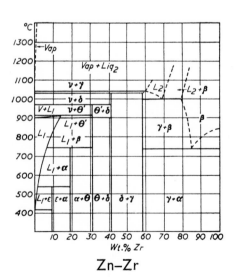

Zn–Zr

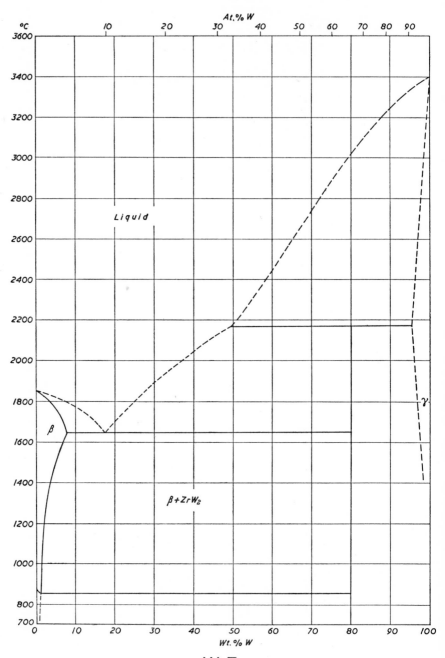

At.% W

°C

Liquid

β

β + ZrW₂

γ

Wt. % W

W–Zr

ACKNOWLEDGEMENTS

These equilibrium diagrams have been taken from the following sources. Acknowledgement and thanks are made to the publishers and authors concerned.

BINARY SYSTEMS

Ag–Al   H. W. L. Phillips, *Inst. Met. Ann. Eq. Diag. No. 21*.
Ag–As   Hansen, plus E. A. Owen and V. W. Rowlands, *J. Inst. Metals*, 1940, **66**, 372.
Ag–Be   H. A. Sloman, *J. Inst. Metals*, 1934, **54**, 166.
Ag–Ce   L. Rolla, A. Iandelli, G. Canneri and R. Vogel, *Z. Metallkunde*, 1943, **35**, 29.
Ag–Cd   F. C. Kracek, " Metals Handbook ", Cleveland, Ohio, 1948.
Ag–Ga   W. Hume-Rothery and K. W. Andrews, *J. Inst. Metals*, 1942, **68**, 137; *idem, Z. Metallkunde*, 1959, **58**, 661.
Ag–Ge   Hansen, plus E. A. Owen and V. W. Rowlands, *J. Inst. Metals*, 1640, **66**, 371.
Ag–Hg   H. M. Day and C. H. Mathewson, *Trans. Amer. Inst. Min. Met. Eng.*, 1938, **128**, 261.
Ag–In   E. Hellner and F. Laves, *Z. Naturforsch*, 1947, A.**2**, 177.
Ag–Li   W. E. Freeth, G. V. Raynor, *J. Inst. Metals*, 1953/4, **82**, 569.
Ag–Mg   W. Hume-Rothery and K. W. Andrews, *J. Inst. Metals*, 1943, **69**, 488; G. F. Sager and B. J. Nelson, " Metals Handbook ", Cleveland, Ohio, 1948; J. L. Haughton and R. J. M. Payne, *J. Inst. Metals*, 1937, **60**, 351.
Ag–Mn   E. Raub, *Z, Metallkunde*, 1949, **40**, (9), 359.
Ag–Pb   A. A. Smith, " Metals Handbook ", Cleveland, Ohio, 1948.
Ag–Sb   Hansen, plus P. W. Reynolds and W. Hume-Rothery, *J. Inst. Metals*, 1937, **60**, 365.
Ag–Sn   Equilibrium Data for Sn Alloys, *Tin Res. Inst.*, 1949.
Ag–Th   E. Raub, *Z. Metallkunde*, 1949, **40**, (11), 431.
Ag–Ti   M. K. McQuillan, *J. Inst. Metals*, 1959/60, **88**, 235.
Ag–Tl   E. Raub, *Z. Metallkunde*, 1949, **40**, (11), 432.
Ag–U   R. W. Buzzard, D. P. Fickle, J. J. Park, *J. Res. Nat. Bur. Stand.*, 1954, **52**, 149.
Ag–Zn   W. Hume-Rothery *et al.*, *Proc. Roy. Soc.*, 1941, A.**177**, 149.
Ag–Zr   J. O. Betterton and D. S. Easton, *Trans. metall. Soc. A.I.M.E.*, 1958, **12**, 470.

Al–As   W. Köster and B. Thoma, *Z. Metallkunde*, 1955, **46**, 291.
Al–Be   H. W. L. Phillips, *Inst. Met. Ann. Diag. No. 19*.
Al–Bi   L. W. Kempf and K. R. van Horn, *Trans. Amer. Inst. Min. Met. Eng.*, 1939, **133**, 85.
Al–Ce   J. H. N. van Gucht, *Z. Metallkunde*, 1957, **48**, 253.
Al–Co   H. W. L. Phillips, *Inst. Met. Ann. Eq. Diag. No. 20*.
Al–Cr   G. Falkenhagen and W. Hofmann, *Z. Metallkunde*, 1950, **41**, 191.
Al–Cu   G. V. Raynor, *Inst. Met. Ann. Diag. No. 4*; R. P. Jewitt and D. J. Mack, *J. Inst. Metals*, 1963–64, **92**, 59.
Al–Fe   H. W. L. Phillips, *Inst. Met. Ann. Diag. No. 13*.
Al–Hf   B. B. Rath, G. P. Mohanty and L. F. Mondolfo, *J. Inst. Metals*, 1960–61, **89**, 248.
Al–In   S. Valentiner and I. Puzicha, *Z. Metallkunde*, 1946, **37**, 127.
Al–Li   K. Schubert, *Z. Metallkunde*, 1950, **41**, (2), 63; W. R. D. Jones and P. P. Das, *J. Inst. Metals*, 1958/59, **87**, 338.
Al–Mg   G. V. Raynor, *Inst. Met. Ann. Diag. No. 5*; J. B. Clark, F. N. Rhines, *T.A.I.M.M.E.*, 1957, **209**, 6.
Al–Mn   H. W. L. Phillips, *J. Inst. Metals*, 1943, **69**, 279.
Al–Mo   F. Sperner, *Z. Metallunde*, 1959, **50**, 58.
Al–Na   C. E. Ransley and H. Neufeld, *J. Inst. Metals*, 1950, **78**, 25.

Al–Ni   W. O. Alexander and N. B. Vaughan, *J. Inst. Metals*, 1937, **61**, 250; H. Groeber and V. Hauk, *Z. Metallkunde*, 1950, **41**, (8), 283; H. W. L. Phillips, *Inst. Met. Ann. Eq. Diag. No. 18.*

Al–Pb   E. Scheil and E. Jahn, *Z. Metallkunde*, 1949, **40**, (8), 319.

Al–Pd   G. Grube and R. Jauch, *Heraeus Festschrift*, 1951, 52.

Al–Sb   H. W. L. Phillips, *Inst. Met. Ann. Eq. Diag. No. 15.*

Al–Si   H. W. L. Phillips, *Inst. Met. Ann. Eq. Diag. No. 16.*

Al–Sn   H. W. L. Phillips, *Inst. Met. Ann. Eq. Diag. No. 14.*

Al–Th   J. R. Murray, *J. Inst. Metals*, 1958–59, **87**, 349.

Al–Ti   E. S. Bumps, H. D. Kessler and M. Hansen, *J. Metals*, 1952, 609, plus W. L. Fink, K. R. van Horn and P. M. Budge, *Trans. Amer. Inst. Min. Met. Eng.*, 1931, **93**, 421; H. W. L. Phillips, *Inst. Met. Ann. Eq. Diag. No. 22*; D. Clarke, K. S. Jepson and G. I. Lewis, *J. Inst. Metals*, 1962/3, **91**, 197.

Al–U   E. Scheil, *Z. Metallkunde*, 1950, **41**, (5), 159.

Al–V   O. N. Carlson, D. J. Kenney, H. A. Wilhelm, *Trans. Amer. Soc. Met.*, 1955, **47**, 520.

Al–W   W. D. Clark, *J. Inst. Metals*, 1940, **66**, 271.

Al–Zn   G. V. Raynor, *Inst. Met. Ann. Diag. No. 1.*

Al–Zr   W. L. Fink and L. A. Willey, *Trans. Amer. Inst. Min. Met. Eng.*, 1939, **133**, 69; D. J. McPherson, M. Hansen, *Trans. Amer. Soc. Met.*, 1954, **46**, 354.

As–Co   Hansen, plus W. Köster and W. Mulfinger, *Z. Metallkunde*, 1938, **30**, 348.

As–Cu   J. C. Mertz and C. H. Mathewson, *Trans. Amer. Inst. Min. Met. Eng.*, 1937, **124**, 69.

As–Ga   W. Köster and B. Thoma, *Z. Metallkunde*, 1955, **46**, 291.

As–In   T. S. Liu and E. A. Peretti, *Trans. Amer. Soc. Met.*, 1953, **45**, 677.

Au–Be   O. Winkler, *Z. Metallkunde*, 1938, **30**, 171.

Au–Bi   W. Earl Lindlief, " Metals Handbook ", Cleveland, Ohio, 1948.

Au–Cd   Hansen, plus E. A. Owen and W. H. Rees, *J. Inst. Metals*, 1941, **67**, 143.

Au–Ce   L. Rolla, A. Iandelli, G. Canneri and R. Vogel, *Z. Metallkunde*, 1943, **35**, 29.

Au–Co   E. Raub and P. Walter, *Z. Metallkunde*, 1950, **41**, 234.

Au–Cu   Hansen, plus H. E. Bennett, *J. Inst. Metals*, 1962/3, **91**, 158.

Au–Ga   Hansen, plus H. Pfisterer, *Z. Metallkunde*, 1950, **41**, 95.

Au–Ge   R. I. Jaffee, E. M. Smith and B. W. Gonser, *Trans. Amer. Inst. Min. Met. Eng.*, 1945, **161**, 366.

Au–Hg   H. M. Day and C. H. Mathewson, *Trans. Amer. Inst. Min. Met. Eng.*, 1938, **128**, 278.

Au–In   Hansen, plus H. Pfisterer, *Z. Metallkunde*, 1950, **41**, (3), 95.

Au–Mn   E. Raub, *Z. Metallkunde*, 1949, **40**, (9), 359; E. Raub, U. Zwicker, H. Bauer, *Z. Metallkunde*, 1953, **44**, 312.

Au–Pt   A. S. Darling, R. A. Mintern and J. C. Chaston, *J. Inst. Metals*, 1952/3, **81**, 125.

Au–Sn   Equilibrium Data for Sn Alloys, *Tin Res. Inst.*, 1949.

Au–Sr   M. Feller-Kniepmaier and T. Heumann, *Z. Metallkunde*, 1960, **51**, 404.

Au–Ti   E. Raub, *Z. Metallkunde*, 1952, **43**, 112; P. Pietrowsky, E. P. Frink, P. Duwez, *T.A.I.M.M.E.*, 1956, **206**, 930.

Au–U   R. W. Buzzard, J. J. Park, *J. Res. Nat. Bur. Stand.*, 1954, **53**, 291.

Au–V   W. Köster and H. Nordskog, *Z. Metallkunde*, 1960, **51**, 501.

Au–Zr   E. Raub and M. Engel, *Z. Metallkunde*, 1948, **39**, 176.

B–Co   W. Köster and W. Mulfinger, *Z. Metallkunde*, 1938, **30**, 348.

B–Mo   R. Steinitz, I. Binder and D. Moskowitz, *J. Metals*, 1952, 983.

B–Nb   H. Novotkny, F. Benesovsky and R. Kieffer, *Z. Metallkunde*, 1959, **50**, 417.

B–Ta   H. Novotkny, F. Benesovsky and R. Kieffer, *Z. Metallkunde*, 1959, **505**, 417.

B–Ti   A. E. Palty, H. Margolin, J. P. Nielsen, *Trans. Amer. Soc. Met.*, 1954, **46**, 312.

B–V   H. Novotkny, F. Benesovsky and R. Kieffer, *Z. Metallkunde*, 1959, **59**, 258.

Ba–Mg   L. A. Carapella, " Metals Handbook ", Cleveland, Ohio, 1948.

Ba–Pb   L. A. Carapella, " Metals Handbook ", Cleveland, Ohio, 1948.

Be–Co   W. Köster and E. Schmid, *Z. Metallkunde*, 1937, **29**, 232, plus G. Venturello and A. Burdese, *Alluminio*, 1951, **20**, (6), 558.

Be–Cr A. R. Edwards, S. T. M. Johnston, *J. Inst. Met.*, 1955/6, **84**, 313.
Be–Fe R. J. Teitel and M. Cohen, *Trans. Amer. Inst. Min. Met. Eng.*, 1949, **185**, 285.
Be–Ni E. Jahn, *Z. Metallkunde*, 1949, **40**, (10), 399.
Be–Pd O. Winkler, *Z. Metallkunde*, 1938, **30**, 162.
Be–U R. W. Buzzard, *J. Res. Nat. Bur. Stand.*, 1953, **50**, 63.

Bi–Co R. Damm, E. Scheil and E. Wachtel, *Z. Metallkunde*, 1962, **53**, 196.
Bi–In O. H. Henry and E. L. Badwick, *Trans. Amer. Inst. Min. Met. Eng.*, 1947, **171**, 389.
Bi–Mn A. U. Seybolt, H. Hansen, B. W. Roberts, P. Yurcisin, *T.A.I.M.M.E.*, 1956, **206**, 606.
Bi–Pb G. O. Hiers, " Metals Handbook ", Cleveland, Ohio, 1948.
Bi–Pd J. Brasier and W. Hume-Rothery, *J. less-common Metals*, 1959, **1**, 157.
Bi–Rh R. G. Goss and W. Hume-Rothery, *J. less-common Metals*, 1962, **4**, 454.
Bi–Sn Equilibrium Data for Sn Alloys, *Tin Res. Inst.*, 1949.
Bi–U P. Cotterill and H. J. Axon, *J. Inst. Metals*, 1958/9, **87**, 159.

C–Cr D. S. Bloom and N. J. Grant, *Trans. Amer. Inst. Min. Met. Eng.*, 1950, **188**, 41.
C–Cu M. B. Bever and C. F. Floe, *Trans. Amer. Inst. Min. Met. Eng.*, 1946, **166**, 128.
C–La F. H. Spedding, K. G. Schneider and A. H. Daane, *Trans. metall. Soc. A.I.M.E.*, 1959, **215**, 192.
C–Mo W. P. Sykes, " Metals Handbook ", Cleveland, Ohio, 1948.
C–Ta F. H. Ellinger, *Trans. Amer. Soc. Metals*, 1943, **31**, 89.
C–Th H. A. Wilhelm and P. Chiotti, *Trans. Amer. Soc. Metals*, 1950, **42**, 1295.
C–Ti H. R. Ogden, R. I. Jaffee, F. C. Holden, *T.A.I.M.M.E.*, 1955, **203**, 73.
C–U H. W. Mallett, A. F. Gerds and H. R. Nelson, *J. Electrochem. Soc.*, 1952, **99**, 197.

Ca–Mg Hansen, plus J. L. Haughton, *J. Inst. Metals*, 1937, **61**, 244.

Cd–Cu Hansen, plus E. Raub, *Metallforsch*, 1947, **2**, 120.
Cd–In S. Valentiner, *Z. Metallkunde*, 1943, **35**, 250; T. Heumann and B. Predil, *Z. Metallkunde*, 1959, **50**, 309.
Cd–Mg W. Hume-Rothery and G. V. Raynor, *Proc. Roy. Soc.*, 1940, A.**174**, 471.
Cd–Sn D. Hansen and W. T. Pell-Walpole, *J. Inst. Metals*, 1936, **59**, 281.
Cd–U A. E. Martin, I. Johnson and H. M. Feder, *Trans. metall. Soc. A.I.M.E.*, 1961, **221**, 789.

Ce–Co R. Vogel, *Z. Metallkunde*, 1946, **37**, 98.
Ce–Fe J. O. Jepson, P. Duwez, *Trans. Amer. Soc. Met.*, 1955, **47**, 543.
Ce–In R. Vogel and H. Klose, *Z. Metallkunde*, 1954, **45**, 633.
Ce–La R. Vogel and H. Klose, *Z. Metallkunde*, 1954, **45**, 633.
Ce–Mg R. Vogel and T. Heumann, *Z. Metallkunde*, 1946, **37**, 1.
Ce–Pb L. Rolla, A. Iandelli and R. Vogel, *Z. Metallkunde*, 1943, **35**, 29.
Ce–Ru W. Obrowski, *Z. Metallkunde*, 1962, **53**, 736.
Ce–Sn Equilibrium Data for Sn Alloys, *Tin Res. Inst.*, 1949.
Ce–Tl L. Rolla, A. Iandelli and R. Vogel, *Z. Metallkunde*, 1943, **35**, 29.
Ce–Th R. T. Weiner, W. E. Freeth, G. V. Raynor, *J. Inst. Metals*, 1957/8, **86**, 185.

Co–Cr A. R. Elsea, A. B. Westerman and G. K. Manning, *Trans. Amer. Inst. Min. Met. Eng.*, 1949, **180**, 579.
Co–Cu C. S. Smith, " Metals Handbook ", Cleveland, Ohio, 1948.
Co–Fe W. C. Ellis and E. S. Greiner, *Trans. Amer. Soc. Metals*, 1941, **29**, 415.
Co–Ga K. Schubert, A. L. Lukas, H. G. Meissner and S. Bhan, *Z. Metallkunde*, 1959, **50**, 534.
Co–Gd V. F. Novy, R. C. Vickery and E. V. Kleber, *Trans. metall. Soc. A.I.M.E.*, 1961, **221**, 588.
Co–Ge H. Pfisterer and K. Schubert, *Z. Metallkunde*, 1949, **40**, 379.
Co–Ir W. Köster and E. Horn, *Z. Metallkunde*, 1952, **43**, 444.
Co–Mn A. Schneider and W. Wunderlich, *Z. Metallkunde*, 1949, **40**, (7), 260.
Co–Mo J. J. Quin and W. Hume-Rothery, *J. less-common Metals*, 1963, **5**, 314.
Co–Nb W. Köster and W. Mulfinger, *Z. Metallkunde*, 1938, **30**, 348.
Co–Ni C. E. Lacy, " Metals Handbook ", Cleveland, Ohio, 1948.
Co–Os W. Köster and E. Horn, *Z. Metallkunde*, 1952, **43**, 444.

Co–Pd   G. Grube and H. Kästner, *Z. Elektrochem.*, 1936, **42**, 156.
Co–Pt   W. Köster, *Z. Metallkunde*, 1949, **40**, (11), 431.
Co–Re   W. Köster and E. Horn, *Z. Metallkunde*, 1952, **43**, 444.
Co–Rh   W. Köster and E. Horn, *Z. Metallkunde*, 1952, **43**, 444.
Co–Ru   W. Köster and E. Horn, *Z. Metallkunde*, 1952, **43**, 444.
Co–Sn   Equilibrium Data for Sn Alloys, *Tin Res. Inst.*, 1949.
Co–Ta   W. Köster and W. Mulfinger, *Z. Metallkunde*, 1938, **30**, 348; M. Karchynsky and R. W. Fountain, *Trans. metall. Soc. A.I.M.E.*, 1959, **215**, 1033.
Co–Ti   F. L. Orrell, M. G. Fontana, *Trans. Amer. Soc. Met.*, 1955, **47**, 554; R. W. Fountain and W. G. Forgang, *Trans. metall. Soc. A.I.M.E.*, 1959, **215**, 998.
Co–V   W. Köster, H. Schmid, *Z. Metallkunde*, 1955, **46**, 195.
Co–W   S. Takeda, *Sci. Rep. Tohoku Imp. Univ.*, 1936. Honda Anniv. Vol., p. 864.
Co–Zn   J. Schramm., *Z. Metallkunde*, 1941, **33**, 46.
Co–Zr   W. Köster and W. Mulfinger, *Z. Metallkunde*, 1938, **30**, 348.

Cr–Fe   A. J. Cook and F. W. Jones, *J. Iron & Steel Inst.*, 1943, **148**, 219.
Cr–Mn   S. J. Carlile, J. W. Christian and W. Hume-Rothery, *J. Inst. Metals*, 1949/50, **76**, 169.
Cr–Mo   O. Kubaschewski and A. Schneider, *Z. Elektrochem.*, 1942, **48**, 671.
Cr–Ni   R. O. Williams, *Trans. Amer. Inst. Min. Met. Eng.*, 1957, **209**, 1257.
Cr–Pd   G. Grube and R. Knabe, *Z. Elektrochem.*, 1936, **42**, 793.
Cr–Pt   E. Gebhardt and W. Köster, *Z. Metallkunde*, 1940, **32**, 262.
Cr–Si   N. N. Kurnakov, *Compt. rend. (Doklady) Acad. Sci. U.S.S.R.*, 1942, **34**, 110.
Cr–Ta   O. Kubaschewski and H. Speidel, *J. Inst. Metals*, 1948/49, **75**, 419.
Cr–Ti   F. B. Cuff, N. J. Grant and C. F. Floe, *J. Metals*, 1952, 848.
Cr–U   A. H. Daane, A. S. Wilson, *T.A.I.M.M.E.*, 1955, **203**, 1219.
Cr–W   W. Trzebiatowski, H. Ploszek and J. Lobzowski, *Analyt. Chem.*, 1947, **19**, 93, plus H. T. Greenaway, *J. Inst. Metals*, 1951/2, **80**, 589.
Cr–Zr   E. T. Hayes, A. H. Roberson and M. H. Davies, *J. Metals*, 1952, 304.

Cu–Ga   W. Hume-Rothery, G. W. Mabbott and K. M. Channel-Evans, *Phil. Trans. Roy. Soc.*, 1934, A.**233**, 1; J. Betterton and W. Hume-Rothery, *J. Inst. Metals*, 1951/2, **80**, 459.
Cu–Ge   W. Hume-Rothery, G. V. Raynor, P. W. Reynolds and H. K. Packer, *J. Inst. Metals*, 1940, **66**, 221; J. Reynolds, W. Hume-Rothery, *J. Inst. Met.*, 1956/7, **85**, 120.
Cu–In   R. O. Jones, E. A. Owen, *J. Inst. Metals*, 1953/4, **82**, 445.
Cu–Mn   B. M. Loring, " Metals Handbook ", Cleveland, Ohio, 1948.
Cu–O   F. N. Rhines and C. H. Mathewson, *Trans. Amer. Inst. Min. Met. Eng.*, 1934, **111**, 339.
Cu–Pb   G. C. Holder, " Metals Handbook ", Cleveland, Ohio, 1939.
Cu–Pd   F. W. Jones and C. Sykes, *J. Inst. Metals*, 1939, **65**, 422.
Cu–S   J. Nutting, *Inst. Met. Ann. Eq. Diag. No.* 24.
Cu–Sb   Hansen, plus J. C. Mertz and C. H. Mathewson, *Trans. Amer. Inst. Min. Met. Eng.*, 1937, **124**, 68.
Cu–Si   C. S. Smith, " Metals Handbook ", Cleveland, Ohio, 1948.
Cu–Sn   G. V. Raynor, *Inst. Met. Ann. Diag. No.* 2.
Cu–Th   E. Raub and M. Engel, *Z. Elektrochem.*, 1943, **49**, 487.
Cu–Ti   A. Joukainen, N. J. Grant and C. F. Floe, *J. Metals*, 1952, 766.
Cu–U   H. A. Wilhelm and O. N. Carlson, *Trans. Amer. Soc. Met.*, 1950, **42**, 1311.
Cu–Y   R. F. Domagela, J. J. Rausch and D. W. Levinson, *Trans. Amer. Soc. Metals*, 1961, **53**, 137.
Cu–Zn   G. V. Raynor, *Inst. Met. Ann. Diag. No.* 3.
Cu–Zr   H. L. Burghoff, " Metals Handbook ", Cleveland, Ohio, 1948.

Fe–Gd   V. F. Novy, R. C. Vickery and E. V. Kleber, *Trans. metall. Soc. A.I.M.E.*, 1961, **221**, 580.
Fe–Ge   R. M. Parke, " Metals Handbook ", Cleveland, Ohio, 1948.
Fe–Mg   A. S. Yue, *J. Inst. Metals*, 1962/3, **91**, 166.
Fe–Mn   A. Hellawell, *Inst. Met. Ann. Eq. Diag. No.* 26.
Fe–Mo   W. J. Gibson, J. R. Lee and W. Hume-Rothery, *J. Iron & Steel Inst.*, 1961, **198**, 64.
Fe–N   K. H. Jack, *Acta. Cryst.*, 1952, **5**, (4), 404.
Fe–Nb   W. J. Gibson, J. R. Lee and W. Hume-Rothery, *J. Iron & Steel Inst.*, 1961, **198**, 64.

Fe–Ni   D. Hanson and J. R. Freeman, *J. Iron & Steel Inst.*, 1923, **107**, 301, plus
         K. Hoselitz, *J. Iron & Steel Inst.*, 1944, **149**, 193.
Fe–Pd   R. M. Parke, " Metals Handbook ", Cleveland, Ohio, 1948.
Fe–Pt   R. M. Parke, " Metals Handbook ", Cleveland, Ohio, 1948.
Fe–Pu   P. G. Marsden, H. R. Haines, J. H. Pearce, M. B. Waldron, *J. Inst. Metals*,
         1957/8, **86**, 166.
Fe–Re   H. Eggers, *Mitt. K. W. Inst. Eisenforsch.*, 1938, **20**, 147.
Fe–Ru   W. S. Gibson and W. Hume-Rothery, *J. Iron & Steel Inst.*, 1958, **189**, 243;
         E. Raub and W. Plate, *Z. Metallkunde*, 1960, **51**, 477.
Fe–S   J. Chipman, " Metals Handbook ", Cleveland, Ohio, 1948.
Fe–Sn   Equilibrium Data for Sn Alloys, *Tin Res. Inst.*, 1949.
Fe–Ta   R. Genders and R. Harrison, *J. Iron & Steel Inst.*, 1936, **134**, 173.
Fe–Ti   G. F. Comstock and J. G. Southand, " Metals Handbook ", Cleveland,
         Ohio, 1948, plus H. W. Worner, *J. Inst. Metals*, 1951, **79**, 173.
Fe–U   J. D. Grogan, *J. Inst. Metals*, 1950, **77**, 571.
Fe–V   A. Hellawell, *Inst. Met. Ann. Eq. Diag. No. 27*.
Fe–Y   R. F. Domagela, J. J. Rausch and D. W. Levinson, *Trans. Amer. Soc.
         Metals*, 1961, **53**, 137.
Fe–Zn   Hansen, plus J. Schramm, *Z. Metallkunde*, 1938, **30**, 131.

Ga–Ge   E. S. Greiner, P. Breidt, *T.A.I.M.M.E.*, 1955, **203**, 187.
Ga–In   J. P. Denny, J. H. Hamilton and J. R. Lewis, *J. Metals*, 1952, 39.
Ga–Mg   H. Gröber and V. Hauk, *Z. Metallkunde*, 1950, **41**, (6), 191.
Ga–Ni   E. Hellner, *Z. Metallkunde*, 1950, **41**, 480.
Ga–Pb   B. Predel, *Z. Metallkunde*, 1959, **50**, 663.
Ga–Pd   K. Schubert, A. L. Lukas, H. G. Meissner and S. Bhan, *Z. Metallkunde*,
         1959, **50**, 534.
Ga–Sb   I. G. Greenfield, R. L. Smith, *T.A.I.M.M.E.*, 1955, **203**, 351.
Ga–Th   B. Predel, *Z. Metallkunde*, 1959, **50**, 663.

Gd–Ni   V. F. Novy, R. C. Vickery and E. V. Kleber, *Trans. metall. Soc. A.I.M.E.*,
         1961, **221**, 585.

Ge–Mg   "Gmelins Handbuch der anorganischen Chemie, Mg. A.4 ", 8th Ed.,
         1952.
Ge–Mn   U. Zwicker, E. Jahn and K. Schubert, *Z. Metallkunde*, 1949, **40**, (12),
         433.
Ge–Ru   E. Raub and W. Fritzsche, *Z. Metallkunde*, 1962, **53**, 779.
Ge–Sn   Equilibrium Data for Sn Alloys, *Tin. Res. Inst.*, 1949.
Ge–Ti   M. K. McQuillan, *J. Inst. Metals*, 1954/5, **83**, 485.
Ge–Zn   E. Gebhardt, *Z. Metallkunde*, 1942, **34**, 255.
Ge–Zr   O. N. Carlson, P. E. Armstrong, H. A. Wilhelm, *Trans. Amer. Soc. Met.*,
         1956, **48**, 843.

H–Ti   G. A. Lenning, C. M. Craighead, R. I. Jaffee, *T.A.I.M.M.E.*, 1954, **200**, 367.
H–Zr   C. E. Ellis, A. D. McQuillan, *J. Inst. Metals*, 1956/7, **85**, 89.

Hf–Mo   A. Taylor, N. J. Doyle and B. J. Kagle, *J. less-common Metals*, 1961, **3**, 265.
Hf–O   E. Rudy and P. Stecher, *J. less-common Metals*, 1963, **5**, 75.
Hf–Re   A. Taylor, B. J. Kagle and N. T. Doyle, *J. less-common Metals*, 1963, **5**, 26.
Hf–U   D. T. Peterson and D. J. Beernstein, *Trans. Amer. Soc. Metals*, 1959, **52**,
         158.
Hf–W   B. C. Giessen, I. Rump and N. J. Grant, *Trans. metall. Soc. A.I.M.E.*,
         1962, **224**, 60; A. Braun and E. Rudy, *Z. Metallkunde*, 1960, **51**, 362.
Hg–In   B. R. Coles, M. F. Merriam and Z. Fisk, *J. less-common Metals*, 1963, **5**, 41.
Hg–Sb   G. Jangg, F. Lihl and E. Legler, *Z. Metallkunde*, 1962, **53**, 313.

Hg–Sn   M. L. Gayler, *J. Inst. Metals*, 1937, **60**, 381.
Hg–U   B. R. T. Frost, *J. Inst. Metals*, 1953/4, **82**, 456.

In–Mg   G. V. Raynor, *Trans. Far. Soc.*, 1948, **44**, 15.
In–Mn   U. Zwicher, *Z. Metallkunde*, 1950, **41**, 400.
In–Ni   E. Hellner, *Z. Metallkunde*, 1950, **41**, 402.
In–Pb   N. Ageev and V. Ageeva, *J. Inst. Metals*, 1936, **59**, 315.
In–Sn   J. C. Blade, E. C. Ellwood, *J. Inst. Metals*, 1956/7, **85**, 30.
In–Ti   R. G. Johnson and R. J. Prosen, *Trans. metall. Soc. A.I.M.E.*, 1962, **224**,
         397.
In–Zn   S. Valentiner, *Z. Metallkunde*, 1943, **35**, 250.

In–Zr   J. O. Betterton and W. K. Noya, *Trans. metall. Soc. A.I.M.E.*, 1958, **212**, 340.

Ir–Mn   E. Raub, W. Mahler, *Z. Metallkunde*, 1955, **46**, 282.

La–Mg   R. Vogel and T. Heumann, *Z. Metallkunde*, 1946, **37**, 1.
La–Ni   R. Vogel, *Z. Metallkunde*, 1946, **37**, 98.
La–Sn   Equilibrium Data for Sn Alloys, *Tin Res. Inst.*, 1949.
La–Tl   L. Rolla, A. Iandelli, G. Canneri and R. Vogel, *Z. Metallkunde*, 1943, **35**, 29.

Li–Mg   W. Hume-Rothery, G. V. Raynor and E. Butchers, *J. Inst. Metals*, 1946, **72**, 538; W. E. Freeth and G. V. Raynor, *J. Inst. Metals*, 1953/4, **82**, 575.

Mg–Mn   W. R. D. Jones, *Inst. Met. Ann. Eq. Diag. No.* 28.
Mg–Pb   Hansen, plus H. Vosskühler, *Z. Metallkunde*, 1939, **31**, 109.
Mg–Pr   L. Rolla, I. Iandelli, G. Canneri and R. Vogel, *Z. Metallkunde*, 1943, **35**, 29.
Mg–Sn   Equilibrium Data for Sn Alloys, *Tin Res. Inst.*, 1949.
Mg–Ti   J. W. Frederekson, *T.A.I.M.M.E.*, 1955, **203**, 368.
Mg–U   P. Chiotti, G. A. Tracey, H. A. Wilhelm, *T.A.I.M.M.E.*, 1956, **206**, 562.
Mg–Y   E. D. Gibson and O. N. Carlson, *Trans. Amer. Soc. Metals*, 1960, **52**, 1084; D. Mizer and J. A. Clark, *Trans. Amer. Inst. Min. Met. Eng.*, 1961, **221**, 207.
Mg–Zn   W. R. D. Jones, *Inst. Met. Ann. Eq. Diag. No.* 29.
Mg–Zr   J. H. Schaum, H. C. Burnett, *J. Res. Nat. Bur. Stand.*, 1952, **49**, 155.

Mn–N   U. Zwicher, *Z. Metallkunde*, 1951, **42**, 274.
Mn–Ni   B. R. Coles and W. Hume-Rothery, *J. Inst. Metals*, 1951/2, **80**, 85.
Mn–P   J. Berak and T. Heumann, *Z. Metallkunde*, 1950, **41**, 21.
Mn–Pd   E. Raub, W. Mahler, *Z. Metallkunde*, 1954, **45**, 430.
Mn–Pt   E. Raub, W. Mahler, *Z. Metallkunde*, 1955, **46**, 282.
Mn–Rh   E. Raub, W. Malher, *Z. Metallkunde*, 1955, **46**, 282.
Mn–Ru   E. Raub, W. Mahler, *Z. Metallkunde*, 1955, **46**, 282
Mn–Sn   Equilibrium Data for Sn Alloys, *Tin Res. Inst.*, **1949**.
Mn–Ti   R. M. Waterstrat, B. N. Das and P. A. Beck, *Trans. metall. Soc. A.I.M.E.*, 1962, **224**, 512.
Mn–U   H. A. Wilhelm and O. N. Carlson, *Trans. Amer. Soc. Met.*, 1950, **42**, 1311.
Mn–V   R. M. Waterstrat, *Trans. metall. Soc. A.I.M.E.*, 1962, **224**, 240.
Mn–Y   R. L. Myklebust and A. H. Daane, *Trans. metall. Soc. A.I.M.E.*, 1962, **224**, 354.
Mn–Zn   E. A. Anderson, " Metals Handbook ", Cleveland, Ohio, 1948.

Mo–Ni   G. Grube and O. Winkler, *Z. Elektrochem.*, 1938, **44**, 427.
Mo–Os   A. Taylor, N. J. Doyle and B. J. Kagle, *J. less-common Metals*, 1962, **4**, 436.
Mo–Pd   C. W. Haworth and W. Hume-Rothery, *J. Inst. Metals*, 1958/9, **82**, 265.
Mo–Re   A. G. Knapton, *J. Inst. Metals*, 1958/9, **87**, 62.
Mo–Rh   C. W. Haworth and W. Hume-Rothery, *J. Inst. Metals*, 1958/9, **87**, 265.
Mo–Ru   E. Anderson and W. Hume-Rothery, *J. less-common Metals*, 1960, **2**, 443.
Mo–Si   R. Kieffer and E. Cerwenka, *Z. Metallkunde*, 1952, **43**, 101.
Mo–Ti   M. Hansen, E. L. Kamen, H. D. Kessler and D. J. McPherson, *J. Metals*, 1951, 881.
Mo–U   P. C. L. Pfeil, *J. Inst. Metals*, 1950, **77**, 553; F. G. Streets and J. J. Stobo, *J. Inst. Metals*, 1963/4, **92**, 171.
Mo–W   W. P. Sykes, " Metals Handbook ", Cleveland, Ohio, 1948.
Mo–Zr   R. F. Domagala, D. J. McPherson and M. Hansen, *Trans. Amer. Inst. Min. Met. Eng.*, 1953, **197**, 73.

N–Ti   A. E. Palty, H. Margolin, J. P. Nielsen, *Trans. Amer. Soc. Met.*, 1954, **46**, 312.

Na–Sn   Equilibrium Data for Sn Alloys, *Tin Res. Inst.*, 1949.
Na–Th   G. Grube and L. Botzenhardt, *Z. Elektrochem.*, 1942, **48**, 418.

Nb–Ni   G. Grube, O. Kubaschewski and K. Zwiauer, *Z. Elektrochem.*, 1939, **45**, 881.
Nb–Re   A. G. Knapton, *J. less-common Metals*, 1959, **1**, 480; B. C. Giessen, R. Nordheim and N. J. Grant, *Trans. metall. Soc. A.I.M.E.*, 1961, **221**, 1009.

Nb–Si   R. Kieffer, F. Benesovsky, H. Schmid, *Z. Metallkunde*, 1956, **47**, 247.
Nb–Th   O. N. Carlson, J. M. Dickinson, H. E. Lunt, H. A. Wilhelm, *T.A.I.M.M.E.*, 1956, **206**, 132.
Nb–Ti   M. Hansen, E. L. Kamen, H. D. Kessler and D. J. McPherson, *J. Metals*, 1951, 881.
Nb–U   P. C. L. Pfeil, J. D. Browne and G. K. Williamson, *J. Inst. Metals*, 1958/9, **87**, 204; B. A. Rogers, E. J. Marthus and M. E. Kirkpatrick, *Trans. metall. Soc. A.I.M.E.*, 1958, **212**, 387.
Nb–V   H. A. Wilhelm, O. N. Carlson, J. M. Dickinson, *T.A.I.M.M.E.*, 1954, **200**, 915.
Nb–Y   C. E. Lundin and D. T. Klodt, *J. Inst. Metals*, 1961/2, **90**, 341.
Nb–Zr   B. A. Rogers, D. F. Atkin, *T.A.I.M.M.E.*, 1955, **203**, 1034.

Ni–Pr   R. Vogel, *Z. Metallkunde*, 1946, **37**, 98.
Ni–Pt   A. Kussmann and H. E. v. Steinwehr, *Z. Metallkunde*, 1949, **40**, (7), 263.
Ni–Si   E. N. Skinner, " Metals Handbook ", Cleveland, Ohio, 1948.
Ni–Sn   K. Schubert and E. Jahn, *Z. Metallkunde*, 1949, **40**, (8), 319.
Ni–Ta   O. Kubaschewski and H. Speidel, *J. Inst. Metals*, 1948/9, **75**, 420.
Ni–Th   L. Horn and C. Bassermann, *Z. Metallkunde*, 1948, **39**, 273.
Ni–Ti   D. M. Poole, W. Hume-Rothery, *J. Inst. Metals*, 1954/5, **82**, 473.
Ni–V   W. B. Pearson and W. Hume-Rothery, *J. Inst. Metals*, 1951/2, **79**, 643.
Ni–Y   B. J. Baudry and A. H. Daane, *Trans. metall. Soc. A.I.M.E.*, 1960, **218**, 854; R. F. Domagela, J. J. Rausch and D. W. Levison, *Trans. Amer. Soc. Metals*, 1961, **53**, 137.
Ni–W   F. H. Ellinger and W. P. Sykes, *Trans. Amer. Soc. Metals*, 1940, **28**, 619.
Ni–Zr   E. T. Hayes, A. H. Roberson, O. G. Paasche, *Trans. Amer. Soc. Met.*, 1953, **45**, 893.
Np–Pu   P. G. Mardon, J. H. Pearce and J. A. C. Marples, *J. less-common Metals*, 1961, **3**, 281.
Np–U   P. G. Mardon, J. H. Pearce and J. A. C. Marples, *J. less-common Metals*, 1959, **1**, 467.

O–Ti   T. H. Schofield, A. E. Bacon, *J. Inst. Metals*, 1955/6, **84**, 47.
O–V   W. Rostoker, A. S. Yamamoto, *Trans. Amer. Soc. Met.*, 1955, **47**, 1002.

Os–W   A. Taylor, B. J. Kagle and N. J. Doyle, *J. less-common Metals*, 1961, **3**, 333.

Pb–Pr   L. Rolla, I. Iandelli, G. Canneri and R. Vogel, *Z. Metallkunde*, 1943, **35**, 29.
Pb–Sn   G. V. Raynor, *Inst. Met. Ann. Diag. No. 6*.
Pb–Te   G. O. Hiers, " Metals Handbook ", Cleveland, Ohio, 1948.
Pb–Ti   P. Farrar, H. Margolin, *T.A.I.M.M.E.*, 1955, **203**, 101.
Pb–U   B. R. T. Frost, J. T. Maskrey, *J. Inst. Metals*, 1953/4, **82**, 171.
Pb–W   S. Inouye, *Mem. Coll. Sci. Kyoto Imp. Univ.*, 1920, **4**, 43.
Pb–Zn   E. A. Anderson and J. L. Rodda, " Metals Handbook ", Cleveland, Ohio, 1939.

Pd–Sn   K. Schubert, A. L. Lukas, H. G. Meissner and S. Bhan, *Z. Metallkunde*, 1959, **50**, 534.
Pd–U   J. A. Catterall, J. D. Grogan, R. J. Pleasance, *J. Inst. Metals*, 1956/7, **85** 63; G. P. Pells, *J. Inst. Metals*, 1963/4, **92**, 416.
Pd–Zn   W. Köster and U. Zwicker, *Hereaus Festschrift*, 1951, 76.
Pd–Zr   K. Anderko, *Z. Metallkunde*, 1959, **60**, 681.

Pr–Sn   Equilibrium Data for Sn Alloys, *Tin Res. Inst.*, 1949.
Pr–Tl   L. Rolla, I. Iandelli, G. Canneri and R. Vogel, *Z. Metallkunde*, 1943, **35**, 29.

Pt–Sn   K. Schubert and E. Jahn, *Z. Metallkunde*, 1949, **40**, (10), 399.
Pt–W   R. I. Jaffee and H. P. Nielsen, *Trans. Amer. Inst. Min. Met. Eng.*, 1949, **180**, 603.
Pt–Zr   E. G. Kendall, C. Hays and R. E. Swift, *Trans. metall. Soc. A.I.M.E.*, 1961, **221**, 445.

Pu–Th   D. M. Poole, G. K. Wilkinson, J. A. C. Maples, *J. Inst. Metals*, 1957/8, **86**, 172.

Re–Ru   E. Rudy, B. Kieffer and H. Fröhlich, *Z. Metallkunde*, 1962, **53**, 90.
Re–W   K. Becker and K. Moers, *Metallwirtschaft*, 1930, **9**, 1063.

Sb–La   R. Vogel and H. Klose, *Z. Metallkunde*, 1954, **45**, 633.
Sb–Sn   E. C. Ellwood, *Inst. Met. Ann. Eq. Diag. No.* 23.
Sb–U   B. J. Baudry and A. H. Daane, *Trans. metall. Soc. A.I.M.E.*, 1959, **215**, 199.
Sb–Zr   J. O. Betterton and W. M. Spicer, *Trans. metall. Soc. A.I.M.E.*, 1958, **212**, 456.

Se–Te   E. Grison, *J. Chem. Phys.*, 1951, **19**, (9), 1109.
Se–Th   R. W. M. D'Eye, P. G. Sellman and J. R. Murray, *J. Chem. Soc.*, 1952, 2555, p. 143; A. Brown and J. J. Norreys, *J. Inst. Metals*, 1960/1, **89**, 238.

Si–Ti   M. Hansen, H. D. Kessler and D. J. McPherson, *Trans. Amer. Soc. Metals*, 1952, **44**, 518.
Si–U   A. Kaufmann, B. Cullity, G. Bitsianes, *T.A.I.M.M.E.*, 1957, **209**, 23.
Si–V   R. Kieffer, F. Benesovsky, H. Schmid, *Z. Metallkunde*, 1956, **47**, 247.
Si–W   R. Kieffer, F. Benesovsky and E. Gallistl, *Z. Metallkunde*, 1952, **43**, 284.
Si–Zr   C. E. Lundin, D. J. McPherson, M. Hansen, *Trans. Amer. Soc. Met.*, 1953, **45**, 901.

Sn–Ti   M. K. McQuillan, *J. Inst. Metals*, 1955/6, **84**, 307.
Sn–Tl   J. C. Blade and E. C. Ellwood, *J. Inst. Metals*, 1959/60, **88**, 186.
Sn–Zn   Equilibrium Data for Sn Alloys, *Tin Res. Inst.*, 1949.
Sn–Zr   D. J. McPherson, M. Hansen, *Trans. Amer. Soc. Met.*, 1953, **45**, 915.

Ta–Th   O. D. MacMaster and W. D. Larsen, *J. less-common Metals*, 1961, **3**, 312.
Ta–Ti   D. Summers Smith, *J. Inst. Metals*, 1952/3, **81**, 73.
Ta–U   C. H. Schramm, P. Gordon and A. R. Kaufmann, *Trans. Amer. Inst. Min. Met. Eng.*, 1950, **188**, 195.
Ta–Y   C. E. Lundin and D. T. Klodt, *J. Inst. Metals*, 1961/2, **90**, 341.
Ta–Zr   D. E. Williams, R. J. Jackson and W. L. Larsen, *Trans. metall. Soc. A.I.M.E.*, 1962, **224**, 751.

Th–Ti   O. N. Carlson, J. M. Dickinson, H. E. Lunt, H. A. Wilhelm, *T.A.I.M.M.E.*, 1956, **206**, 132.
Th–U   J. R. Murray, *J. Inst. Metals*, 1958/9, **87**, 94.
Th–Zn   P.Chiotti and K. J. Gill, *Trans. metall. Soc. A.I.M.E.*, 1961, **221**, 573.

Ti–U   A. G. Knapton, *J. Inst. Metals*, 1954/5, **83**, 497.
Ti–V   H. N. Aderstedt, T. R. Pequignot and T. M. Rayner, *J. Amer. Chem. Soc.*, 1952, **44**, 990.
Ti–Y   D. W. Bau, *Trans. Amer. Soc. Metals*, 1961, **53**, 1.
Ti–Zn   W. Heine and U. Zwicker, *Z. Metallkunde*, 1962, **53**, 380.

U–V   H. A. Saller, F. A. Rough, *J. Metals*, 1953, 545.
U–W   C. H. Schramm, P. Gordon and A. R. Kaufmann, *Trans. Amer. Inst. Min. Met. Eng.*, 1950, **188**, 195.
U–Zn   P. Chiotti, H. H. Klepfer, K. J. Gill, *T.A.I.M.M.E.*, 1957, **209**, 51.
U–Zr   D. Summers Smith, *J. Inst. Metals*, 1954/5, **82**, 277.

V–Y   C. E. Lundin and D. T. Klodt, *J. Inst. Metals*, 1961/2, **90**, 341.
V–Zr   J. T. Williams, *T.A.I.M.M.E.*, 1955, **203**, 345.

W–Zr   R. F. Domagala, D. J. McPherson and M. Hansen, *Trans. Amer. Inst. Min. Met. Eng.*, 1953, **197**, 73.

Zn–Zr   P. Chiotti and G. R. Kilp, *Trans. metall. Soc. A.I.M.E.*, 1959, **215**, 892.

TERNARY SYSTEMS

(No diagrams for these systems have been included)

Ag–Al–Cu	C. Panseri and M. Leoni, *Alluminio*, 1961, **306**, 289.
Ag–Al–Mg	J. L. Haughton, *J. Inst. Metals*, 1939, **65**, 447.
Ag–Al–Mn	D. W. Wakeman and G. V. Raynor, *J. Inst. Metals*, 1948/9, **75**, 131.
Ag–Al–Sn	C. S. Cheng, S. J. Huang, S. I. Chen, I. M. Chen and K. S. Kan, *Acta phys. sin.*, 1958. **14**, 346.
Ag–Al–Ti	W. Köster and A. Sampaio, *Z. Metallkunde*, 1957, **48**, 331.
Ag–Al–Zn	H. Haneman and A. Schrader, "Ternäre Legierungen des Aluminiums", Dusseldorf, 1952, p. 56.
Ag–Au–Cu	*International Critical Tables*, 1927, **2**, 446.
Ag–Au–Ni	*International Critical Tables*, 1927, **2**, 442.
Ag–Au–Pd	E. M. Wise, "Metals Handbook", Cleveland, Ohio, 1948.
Ag–Bi–Zn	E. Henglein and W. Köster, *Z. Metallkunde*, 1948, **39**, 391.
Ag–Cd–Cu	E. Gebhart and G. Petzow, *Z. Metallkunde*, 1956, **47**, 401.
Ag–Cd–Sn	E. Gebhardt and G. Petzow, *Z. Metallkunde*, 1959, **50**, 696.
Ag–Cd–Zn	G. Petzow and E. Wagner, *Z. Metallkunde*, 1962, **53**, 189.
Ag–Cu–In	E. Gebhardt and M. Dreher, *Z. Metallkunde*, 1951, **42**, 230, and 1952, **43**, 357.
Ag–Cu–Pd	E. M. Wise, "Metals Handbook", Cleveland, Ohio, 1948.
Ag–Cu–Sn	E. Gebhardt and G. Petzow, *Z. Metallkunde*, 1959, **50**, 597.
Ag–Cu–Zn	R. H. Leach, "Metals Handbook", Cleveland, Ohio, 1948; E. Gebhardt, G. Petzow and W. Krauss, *Z. Metallkunde*, 1962, **53**, 461.
Ag–Hg–Sn	M. L. V. Gayler, *J. Inst. Metals*, 1937, **60**, 379.
Ag–Li–Mg	W. Hume-Rothery, G. V. Raynor and E. Butchers, *J. Inst. Metals*, 1945, **71**, 589.
Ag–Mg–Sb	B. R. T. Frost and G. V. Raynor, *Proc. Roy. Soc.*, 1950, A**203**, (1072), 132.
Ag–Mg–Sn	G. V. Raynor and B. R. T. Frost, *J. Inst. Metals*, 1948/9, **75**, 777.
Ag–Mg–Zn	G. V. Raynor and R. A. Smith, *J. Inst. Metals*, 1949/50, **76**, 389.
Ag–Pb–Sb	B. Blumenthal, *Trans. Amer. Inst. Min. Met. Eng.*, 1944, **156**, 241.
Ag–Pb–Zn	W. Seith and G. Helmold, *Z. Metallkunde*, 1951, **42**, (5), 138.
Ag–Pd–Pt	R. H. Leach, "Metals Handbook", Cleveland, Ohio, 1948.
Al–B–Co	H. H. Stadelmeier and R. A. Gregg, *Metall.*, 1962, **16**, 405.
Al–B–Ni	H. H. Stadelmeier and A. C. Fraker, *Z. Metallkunde*, 1962, **53**, 214.
Al–Be–Mg	H. Haneman and A. Schrader, "Ternäre Legierungen des Aluminiums", Dusseldorf, 1952.
Al–Be–Si	H. Haneman and A. Schrader, "Ternäre Legierungen des Aluminiums", Dusseldorf, 1952.
Al–Bi–Mg	H. Haneman and A. Schrader, "Ternäre Legierungen des Aluminiums", Dusseldorf, 1952.

Al–C–Fe		"Gmelin's Handbuch der anorganischen Chemie, Al A.8", 8th Ed., 1950.
Al–Ca–Mg		Z. A. Catterall and R. J. Pleasance, *J. Inst. Metals*, 1957/8, **86**, 189.
Al–Ca–Si		H. Haneman and A. Schrader, "Ternäre Legierungen des Aluminiums", Dusseldorf, 1952.
Al–Cd–Mg		E. Jäneche, *Z. Metallkunde*, 1938, **30**, (4), 424.
Al–Cd–Mn		G. V. Raynor and D. W. Wakeman, *Phil. Mag.*, 1948, VII, **39**, 245.
Al–Cd–Sn		A. J. Perry and H. J. Bray, *J. Inst. Metals*, 1963/4, **92**, 152.
Al–Cd–Zn		*International Critical Tables*, 1927, **2**, 409.
Al–Co–Cu		P. C. L. Pfeil and G. V. Raynor, *Proc. Roy. Soc.*, 1949, A.**197**, (1050), 321.
Al–Co–Fe		G. V. Raynor and M. B. Waldron, *Proc. Roy. Soc.*, 1948, A.**194**, (1038), 362.
Al–Co–Mn		G. V. Raynor, *J. Inst. Metals*, 1947, **73**, 521.
Al–Co–Ni		G. V. Raynor and P. C. L. Pfeil, *J. Inst. Metals*, 1947, **73**, 609.
Al–Cr–Fe		J. N. Pratt and G. V. Raynor, *J. Inst. Metals*, 1951/2, **80**, 449.
Al–Cr–Mg	Liquidus	H. Haneman and A. Schrader, "Ternäre Legierungen des Aluminiums", Dusseldorf, 1952.
Al–Cr–Mn		G. V. Raynor and K. Little, *J. Inst. Metals*, 1945, **71**, 493; J. W. H. Clare, *Trans. metall. Soc. A.I.M.E.*, 1959, **215**, 429.
Al–Cr–Tl		E. Ence, P. A. Farrar and H. Marjolin, *Wright Air Corpn. Dev. Div., Tech. Rep.*, 1960, **60**.
Al–Cr–V		E. Ence, P. A. Farrar and H. Marjolin, *Wright Air Corpn. Dev. Div. Tech. Rep.*, 1960, **60**.
Al–Cr–Zn		A. R. Harding and G. V. Raynor, *J. Inst. Metals*, 1951/2, **80**, 435, plus K. Little, H. J. Axon and W. Hume-Rothery, *J. Inst. Metals*, 1948/9, **75**, 39.
Al–Cu–Fe	Liquidus	H. W. L. Phillips, *Inst. Met. Mon. Rep. Series*, no. 25.
	Phases in solid	A. J. Bradley and H. J. Goldschmidt, *J. Inst. Metals*, 1939, **65**, 391.
Al–Cu–In		P. H. Stirling and G. V. Raynor, *J. Inst. Metals*, 1955/6, **84**, 57.
Al–Cu–Li		H. K. Hardy and J. M. Silcox, *J. Inst. Metals*, 1955/6, **84**, 423.
Al–Cu–Mg	Liquidus Al rich	H. W. L. Phillips and N. S. Brommelle, *J. Inst. Metals*, 1949, **75**, 529.
	430° Isothermal Al rich	H. W. L. Phillips and N. S. Brommelle, *J. Inst. Metals*, 1949, **75**, 529.
	Liquidus Mg rich	A. Beck, Mg und seiner Legierungen, Springer, Berlin, 1939, Fig. 110.
	460° Isothermal	A. T. Little, W. Hume-Rothery and G. V. Raynor, *J. Inst. Metals*, 1944, **70**, 491, plus D. J. Strawbridge, W. Hume-Rothery and A. T. Little, *J. Inst. Metals*, 1948, **74**, 198, plus H. W. L. Phillips, *Inst. Met. Mon. Rep. Series*, no. 25.
Al–Cu–Mn	Liquidus Al rich	H. W. L. Phillips and M. K. B. Day, *J. Inst. Metals*, 1948, **74**, 33, plus D. R. F. West and D. L. Thomas, *J. Inst. Metals*, 1956/7, **85**, 97, plus H. W. L. Phillips, *Inst. Metals*, Mon. *Rep. Series*, no. 25.
	500° Isothermal Al rich	H. G. Petri, *Aluminium Archiv.*, 1938, (**14**).
	500°, 800° Isothermal Cu rich	R. S. Dean, J. R. Long, T. A. Graham, A. H. Roberson and C. E. Armantrout, *Trans. Amer. Inst. Min. Met. Eng.*, 1947, **171**, 70.

Al–Cu–Ni	Phases Room Temp.	A. J. Bradley, W. L. Bragg and C. Sykes, *J. Iron & Steel Inst.*, 1940, **141**, 104.
	Liquidus	W. Köster, U. Zwicker and K. Moeller, *Z. Metallkunde*, 1948, **39**, 225.
Al–Cu–Si	Liquidus Al rich	H. W. L. Phillips, *Inst. Met. Mon. Rep. Series*, no. 25.
	500° Isothermal	H. Wiehr, *Aluminium Archiv.*, 1940, 31.
	Section diagram	F. H. Wilson, *Trans. Amer. Inst. Min. Met. Eng.*, 1948, **175**, 262.
Al–Cu–Sn	Liquidus	*International Critical Tables*, 1927, **2**, 408.
Al–Cu–Ti		C. Panseri and M. Leoni, *Alluminio*, 1962, **31**, 461; V. N. Vigdorovich, M. V. Maltsev and A. N. Kristovnikow, *Izv. vysch. uches. Zaved.*, 1958, **2**, 142.
Al–Cu–Zn	460° Isothermal Al rich	D. J. Strawbridge, W. Hume-Rothery and A. T. Little, *J. Inst. Metals*, 1948, **74**, 191.
	Liquidus	*International Critical Tables*, 1927, **2**, 407.
	300° Isothermal Zn rich	*International Critical Tables*, 1927, **2**, 408.
Al–Fe–Mg	Liquidus	H. W. L. Phillips, *J. Inst. Metals*, 1946, **72**, 173, plus H. W. L. Phillips, *Inst. Met. Mon. Rep. Series*, no. 25.
Al–Fe–Mn	Liquidus	H. W. L. Phillips, *J. Inst. Metals*, 1943, **69**, 275, plus H. W. L. Phillips, *Inst. Met. Mon. Rep. Series*, no. 25.
	600° Isothermal	E. Degischer, *Aluminium Archiv.*, 1939, **18**.
Al–Fe–Ni	Phases	A. J. Bradley and A. Taylor, *J. Inst. Metals*, 1940, **66**, 58.
	Liquidus	H. W. L. Phillips, *J. Inst. Metals*, 1942, **68**, 37, plus H. W. L. Phillips, *Inst. Met. Mon. Rep. Series*, no. 25.
	Liquidus	W. Köster, *Arch. Eisenhüttenw.*, 1933, **7**, 257.
	3 Sectionals	W. Köster, *Arch. Eisenhüttenw.*, 1933, **7**, 257.
Al–Fe–Si	Liquidus	H. W. L. Phillips, *J. Inst. Metals*, 1946, **72**, 173, plus H. W. L. Phillips, *Inst. Met. Mon. Rep. Series*, no. 25.
	Liquidus	"Gmelin's Handbuch der anorganischen Chemie, Al A.8 ", 8th Ed., 1950.
Al–Fe–Ti		Y. V. Bok and M. V. Maltsev, *Izv. vyssh. ucheb. Zaved.*, 1948, **3**, 110.
Al–Fe–Zn		E. Gebhart, *Z. Metallkunde*, 1953, **44**, 206.
Al–In–Sn		A. N. Campbell, L. B. Buchanan, J. M. Kuzmak and R. H. Tuxworth, *J. Amer. Chem. Soc.*, 1952, **74**, (8), 1962.
Al–Li–Mg		J. A. Rowland, C. E. Armantrout, and D. F. Walsh. *T.A.I.M.M.E.*, 1955, **203**, 355.
Al–Mg–Mn	Liquidus	W. G. Leeman, *Aluminium Archiv.*, 1938, (**9**).
	400° Isothermal	D. W. Wakeman and G. V. Raynor, *J. Inst. Metals*, 1948/49, **75**, 131.
	Liquidus Al corner	E. Butchers, G. V. Raynor and W. Hume-Rothery, *J. Inst. Metals*, 1943, **69**, 217.
Al–Mg–Ni	Liquidus	H. Haneman and A. Schrader, "Ternäre Legierungen des Aluminiums", Dusseldorf, 1952.
Al–Mg–Pb	Liquidus	H. Haneman and A. Schrader, "Ternäre Legierungen des Aluminiums", Dusseldorf, 1952.
Al–Mg–Si	Liquidus Al rich	H. W. L. Phillips, *J. Inst. Metals*, 1946, **72**, 173, plus H. W. L. Phillips, *Inst. Met. Mon. Rep. Series*, no. 25.
	400° Isothermal	H. W. L. Phillips, *J. Inst. Metals*, 1946, **72**, 157.
	Liquidus Mg rich	A. Beck, Mg. u.s. Legierungen, Springer, Berlin, 1939, Fig. 116.
Al–Mg–Zn	Liquidus Al rich	E. Butchers, G. V. Raynor and W. Hume-Rothery, *J. Inst. Metals*, 1943, **69**, 214, plus H. W. L. Phillips, *Inst. Met. Mon. Rep. Series*, no. 25.
	Phases, solid	E. Butchers, G. V. Raynor and W. Hume-Rothery, *J. Inst. Metals*, 1943, **69**, 214.

	Liquidus Mg rich	A. Beck, Mg. u.s. Legierungen, Springer, Berlin 1939, Fig. 119.
	460° Isothermal	D. J. Strawbridge, W. Hume-Rothery and A. T. Little, *J. Inst. Metals*, 1948, **74**, 191 H. Watanabe, *Nippon Kink. Gakk.*, 1959, **23**, 285.
Al–Mn–Ni		G. V. Raynor, *J. Inst. Metals*, 1944, **70**, 509
Al–Mn–Si		H. W. L. Phillips, *J. Inst. Metals*, 1943,. **69** 297, plus H. W. L. Phillips, *Inst. Met. Mon Rep. Series*, no. 25.
Al–Mn–Ti		R. F. Domagala and W. Rostoker, *Trans. Amer Soc. Met.*, 1955, **47**, 565; M. V. Maltsev and Y. V. Bok, *Izv. vyssh. ucheb. Zaved.*, 1958 **3**, 138; T. Sato, Y. C. Yuang and Y. Kondo *Sumitomo lt. Metal tech., Rep.* 1960, **1**, 36.
Al–Mn–Zn	Liquidus Al corner	E. Butchers, G. V. Raynor and W. Hume Rothery, *J. Inst. Metals*, 1943, **69**, 217.
	600° Isothermal	G. V. Raynor and D. W. Wakeman, *Phil. Mag.* 1948, VII, **39**, 245.
Al–Mo–Ni		R. W. Guard and E. A. Smith, *J. Inst. Metals* 1959/60, **88**, 283.
Al–Mo–V		F. Sperner, *Z. Metallkunde*, 1959, **50**, 592.
Al–Na–Si		C. E. Ransley and H. Neufeld, *J. Inst. Metals*, 1950, **78**, 25.
Al–Ni–Si	Liquidus	H. W. L. Phillips, *J. Inst. Metals*, 1942, **68**, 27 R. W. Guard and E. A. Smith, *J. Inst. Metals*, 1959/60, **88**, 369.
Al–Ni–Ti	1150° & 750° Isothermal	A. Taylor and R. W. Floyd, *J. Inst. Metals*, 1952/3, **81**, 25.
Al–Sb–Si	Liquidus	H. Haneman and A. Schrader, "Ternäre Legierungen des Aluminiums", Dusseldorf, 1952.
Al–Si–U		G. Petzow and I. Kvernes, *Z. Metallkunde*, 1962, **53**, 248.
Al–Si–V		E. Gebhardt and G. Joseph, *Z. Metallkunde*, 1961, **52**, 310.
Al–Ti–Si		F. H. Crossley and D. H. Turner, *Trans. metall. Soc. A.I.M.E.*, 1958, **212**, 60.
Al–Ti–V		C. B. Jordan and P. Duwez, *Trans. Amer. Soc. Met.*, 1956, **48**, 783; P. A. Farrar and A. Marjolin, *Trans. metall. Soc. A.I.M.E.*, 1961, **221**, 1214.
As–In–Sb		C. H. Shih and E. A. Peretti, *Trans. Amer. Soc. Met.*, 1956, **48**, 706.
As–Ga–Zn		W. Köster and W. Ulrich, *Z. Metallkunde*, 1958, **49**, 361.
Au–Fe–Ni		W. Köster and W. Abrici, *Z. Metallkunde*, 1961, **52**, 383.
B–Ni–Sn		H. H. Stadelmeier and L. T. Jordan, *Z. Metallkunde*, 1962, **53**, 11, 719.
B–Ni–Zn		H. H. Stadelmeier, J. D. Schobul and L. T. Jordan, *Metals*, 1962, **16**, 752.
Be–Cu–Mn		M. V. Maltsev and C. Shi-Chan, *Tsvet. Metally. N.Y.*, 1960, **1**, 138.
Be–Cu–Sn	700° Isothermal	R. A. Cresswell and J. W. Cuthbertson, *Trans. Amer. Inst. Min. Met. Eng.*, 1951, **191**, 782.
Be–Fe–Si		R. Vogel and H. J. Geste, *Ard. Eisenhüttenw.*, 1960, **31**, 319.
Bi–Cd–In		E. A. Peretti, *Amer. Soc. Metals, Preprint*, 1960, **184**.
Bi–Cd–Pb	Liquidus	Teh. Hsuan Ho, W. Hofmann and H. Hanemann, *Z. Metallkunde*, 1953, **44**, 127.
Bi–Cd–Sn		H. J. Bray, F. D. Bell and S. J. Harris, *J. Inst. Metals*, 1961/2, **90**, 24.
Bi–Cd–Zn	Solubility	*International Critical Tables*, 1927, **2**, 448.

Bi–Cu–Zn	Liquidus	E. Henglein and W. Köster, *Z. Metallkunde*, 1948, **39**, 391.
Bi–Mg–Pb		Y. K. Silina and V. D. Ponomarev, *Tsvet. Metally*, N.Y., 1960, **1**, 91.
Bi–Mg–Zn	Liquidus	"Gmelin's Handbuch der anorganischen Chemie, Mg A.4", 8th Ed., 1952.
Bi–Pb–Sn	Liquidus	Teh-Hsuan Ho, W. Hofmann and H. Hanemann, *Z. Metallkunde*, 1953, **44**, 127.
Bi–Pb–Zn	Liquidus	S. D. Muzaffar and R. Chand, *J. Amer. Chem. Soc.*, 1944, **66**, 1374.
Bi–Sn–Zn	Liquidus	*International Critical Tables*, 1927, **2**, 418.
C–Co–Sn		J. L. Hütter, H. H. Stadelmeier and W. K. Hardy, *Trans. metall. Soc. A.I.M.E.*, 1960, **218**, 859.
C–Co–W		P. Rautala and J. T. Norton, *Trans. Amer. Inst. Min. Met. Eng.*, 1952, **194**, 1045.
C–Cr–Fe		W. Crafts, "Metals Handbook", Cleveland, Ohio, 1939; N. R. Griffing, W. D. Forgeng and G. W. Healey, *Trans. metall. Soc. A.I.M.E.*, 1962, **224**, 148.
C–Fe–Mn		C. Wells, "Metals Handbook", Cleveland, Ohio, 1939.
C–Fe–Mo		C. Wells, "Metals Handbook", Cleveland, Ohio, 1939.
C–Fe–Ni		J. H. Andrew, G. T. C. Bottomley, W. R. Maddocks and R. T. Percival, *J. Iron & Steel Inst. Sp. Rept.*, 1938, 23.
C–Fe–P		*International Critical Tables*, 1927, **2**, 454.
C–Fe–Si		J. E. Hilliard and W. S. Owen, *J. Iron & Steel Inst.*, 1952, **172**, 268.
C–Fe–Ti		M. F. Hawkes, "Metals Handbook", Cleveland, Ohio, 1948.
C–Fe–V		F. Wever, A. Rose and H. Eggers, *Mitt. k.w. Inst. Eisenf.*, 1936, **18**, 239.
C–Fe–W		G. A. Roberts and A. H. Grobe, "Metals Handbook", Cleveland, Ohio, 1948.
C–In–Ni		L. J. Hutter, H. H. Stadelmeier and A. C. Fraker, *Metall.*, 1960, **14**, 29.
C–Mo–U		F. G. Streets and J. J. Stob, *J. Inst. Metals*, 1963/4, **92**, 171.
C–Ni–Ti		E. R. Stover and J. Wulft, *Trans. metall. Soc. A.I.M.E.*, 1959, **215**, 127.
C–V–W		E. Rudy, F. Benesowski, and E. Rudy, *Montash. Chem.*, 1962, **93**, 693.
Ca–Mg–Zn	Liquidus	"Gmelin's Handbuch der anorganischen Chemie, Mg A.4", 8th Ed., 1952; J. B. Clark, *Trans. metall. Soc. A.I.M.E.*, 1961, **221**, 644.
Cd–Cu–Sn		E. Gebhardt and G. Petzov, *Z. Metallkunde*, 1959, **50**, 668.
Cd–Ga–Pb		B. Predil, *Z. Metallkunde*, 1961, **52**, 507.
Cd–In–Sn	Liquidus	E. Gebhardt, *Z. Metallkunde*, 1949, **40**, (12), 438.
Cd–In–Zn		S. C. Carapella and E. A. Peretti, *Trans. Amer. Soc. Metals*, 1951, **43**, 853.
Cd–Mg–Pb	Liquidus	E. Jänecke, *Z. Metallkunde*, 1938, **30**, 395.
Cd–Mg–Zn	Liquidus	E. Jänecke, *Z. Metallkunde*, 1938, **30**, 424.
Cd–Pb–Sb		E. C. Rollason and V. B. Hysel, *J. Inst. Metals*, 1940, **66**, 349.
Cd–Pb–Sn	Liquidus	A. Stoffel, *Z. anorg. allg. Chem.*, 1907, **53**, 137.
Cd–Pb–Zn	Liquidus	M. Cook, *J. Inst. Metals*, 1924, **31**, 297.
Cd–Sb–Sn		D. Hanson and W. T. Pell Walpole, *J. Inst. Metals*, 1937, **61**, 265.
Cd–Sn–Tl	Liquidus	E. Jänecke, *Z. Metallkunde*, 1939, **31**, 170.
Cd–Sn–Zn	Liquidus	R. Lorenz and D. Plumbridge, *Z. anorg. allg. Chem.*, 1913, **83**, 228; H. J. Brar, *J. Inst. Metals*, 1958/9, **87**, 49.

Co–Cr–Fe	1200° Isothermal	S. Rideout, W. D. Manly, E. L. Kamen, B. S. Lement and P. A. Beck, *Trans. Amer. Inst. Min. Met. Eng.*, 1951, **191**, 872; W. Köster and G. Hofmann, *Arch. Eisenhüttenw.*, 1959, **30**, 249.
Co–Cr–Mo	1200° Isothermal	S. Rideout, W. D. Manly, E. L. Kamen, B. S. Lement and P. A. Beck, *Trans. Amer. Inst. Min. Met. Eng.*, 1951, **191**, 872.
Co–Cr–Ni	1200° Isothermal	S. Rideout, W. D. Manly, E. L. Kamen, B. S. Lement and P. A. Beck, *Trans. Amer. Inst. Min. Met. Eng.*, 1951, **191**, 872.
Co–Cr–Ti		E. K. Zakanhov and B. J. Livshits, *Izv. Akad. Nauk. S.S.S.R.*, 1962, **5**, 143.
Co–Fe–Mo	1200° Isothermal	D. K. Das, S. P. Rideout and P. A. Beck, *Trans. Amer. Inst. Min. Met. Eng.*, 1952, **194**, 1071.
Co–Fe–Ni	Phases, solid	A. J. Bradley, W. L. Bragg and C. Sykes, *J. Iron & Steel Inst.*, 1940, **141**, 109.
Co–Fe–Pd		V. V. Cuprina and A. T. Grigorev, *Zh. neorg. Khim.*, 1958, **3**, 2736.
Co–Fe–V		D. L. Martin and A. H. Geisler, *Trans. Amer. Soc. Metals*, 1952, **44**, 461.
Co–Ni–Mo	1200° Isothermal	D. K. Das, S. P. Rideout and P. A. Beck, *Trans. Amer. Inst. Min. Met. Eng.*, 1952, **194**, 1071.
Co–Ni–Zn		W. Köster, H. Schmid, and E. Dahesh, *Z. Metallkunde*, 1956, **47**, 165.
Cr–Cu–Ni		*International Critical Tables*, 1927, **2**, 443.
Cr–Cu–Zr		M. V. Zakharov, M. V. Stepanove and V. M. Glazov, *Metallov. Obrab. Metall.*, 1957, **3**, 23.
Cr–Fe–Mn		G. Masing and R. Vogel, "Handbuch der Metal Physik", vol. 2, Leipsig, 1937.
Cr–Fe–Mo		J. N. Putman, R. D. Potter and N. J. Grant, *Trans. Amer. Soc. Metals*, 1951, **43**, 834; J. G. McMullin, S. F. Reiter and D. G. Ebeling, *Trans. Amer. Soc. Met.*, 1954, **46**, 799.
Cr–Fe–Ni		E. C. Bain and R. H. Aborn, "Metals Handbook", Cleveland, Ohio, 1948.
	Liquidus	C. H. N. Jenkins, E. H. Bucknell, C. R. Austin and G. A. Mellor, *J. Iron & Steel Inst.*, 1937, **136**, 187.
	650° Isothermal	E. C. Bain and R. H. Aborn, "Metals Handbook", Cleveland, Ohio, 1948.
	Phases in solid	A. J. Bradley, W. L. Bragg and C. Sykes, *J. Iron & Steel Inst.*, 1940, **141**, 108.
Cr–Fe–Ti		R. J. van Thyne, H. D. Kessler and M. Hansen, *J. Metals*, 1953, 1209; A. G. Boriskine and I. I. Kornilov, *Izv. Akad. Nauk, S.S.S.R.*, 1961, **3**, 34.
Cr–Mo–Nb		M. I. Zakarova, D. A. Prokoshkin, *Izv. Akad. Nauk., S.S.S.R.*, 1961, **4**, 59; H. J. Goldschmidt and J. A. Brand, *J. less-common Metals*, 1961, **3**, 44.
Cr–Mo–Ni	1200° Isothermal	S. Rideout, W. D. Manly, E. L. Kamen and P. A. Beck, *Trans. Amer. Inst. Min. Met. Eng.*, 1951, **191**, 872.
Cr–Nb–Si		H. J. Goldschmidt and J. A. Brand, *J. less-common Metals*, 1961, **3**, 34.
Cr–Ni–Si		R. W. Guard and E. A. Smith, *J. Inst. Metals*, 1959/60, **88**, 369.
Cr–Ni–Ti	1150° & 750° Isothermals	A. Taylor and R. W. Floyd, *J. Inst. Metals*, 1951/2, **80**, 577.
Cr–Ni–W		I. I. Kornilov and G. P. Budberg, *Trudy Inst. Metall. A.A. Baikova*, 1957, **1**, 132.
Cr–O–Ti		W. Rostowker, *T.A.I.M.M.E.*, 1955, **203**, 113.
Cu–Au–Pd		E. Raub, G. Worwag, *Z. Metallkunde*, 1955, **46**, 119.

Cu–Fe–Ni	500°–900° Isothermals	E. W. Palmer and F. H. Wilson, *J. Metals*, 1952, 55.
Cu–Fe–Si		A. G. H. Andersen and A. W. Kingsbury, *Trans. Amer. Inst. Min. Met. Eng.*, 1943, **152**, 38.
Cu–Ga–Ge		G. V. Raynor and T. B. Massalski, *J. Inst. Metals*, 1955/6, **84**, 66.
Cu–Ge–Zn		P. Greenfield and G. V. Raynor, *J. Inst. Metals*, 1951/2, **80**, 375.
Cu–Mg–Ni	Liquidus	W. Köster, *Z. Metallkunde*, 1951, **42**, 327.
Cu–Mg–Sb	Liquidus	"Gmelin's Handbuch der anorganischen Chemie", Mg A.4, 8th Ed., 1952; R. Dobbener and R. Vogel, *Z. Metallkunde*, 1959, **50**, 412.
Cu–Mg–Zn	Liquidus	W. Köster, *Z. Metallkunde*, 1948, **39**, 352.
Cu–Mn–Ni	Liquidus	*International Critical Tables*, 1927, **2**, 445.
	Solubility	*International Critical Tables*, 1927, **2**, 446.
	700° & 500° Isothermals	U. Zwicker, *Z. Metallkunde*, 1951, **42**, 331.
Cu–Mn–Si		C. S. Smith and W. R. Hibbard, *Trans. Amer. Inst. Min. Met. Eng.*, 1942, **147**, 222, plus C. W. Funk and J. A. Rowland, *J. Metals*, 1953, 723.
Cu–Mn–Zn		T. R. Graham, J. R. Long, C. E. Armantrout and A. H. Roberson, *Trans. Amer. Inst. Min. Met Eng.*, 1949, **185**, 675.
Cu–Ni–Sb		N. Sibata, *Sc. Rep. Imp. Univ. Tohoku*, 1942, **1**, 30, **2**, 154, 193.
Cu–Ni–Sn		J. T. Eash, "Metals Handbook", Cleveland, Ohio, 1939.
Cu–Ni–Zn		J. Schramm, "Metals Handbook", Cleveland, Ohio, 1939.
Cu–Pb–Zn	Phases in solid	*International Critical Tables*, 1927, **2**, 418.
	Liquidus	E. Henglein and W. Köster, *Z. Metallkunde*, 1948, **39**, 398.
Cu–Pd–Ag		E. Raub and G. Wörwag, *Z. Metallkunde*, 1955, **46**, 52.
Cu–Sb–Sn		J. V. Harding and W. T. Pell Walpole, *J. Inst. Metals*, 1948/9, **75**, 124.
Cu–Si–Zn		G. Mime and M. Hasegawa, *Nippon kink. Gakk.*, 1959, **23**, 585.
Cu–Sn–Zn	Liquidus	G. Tammann and M. Hansen, "Metals Handbook", Cleveland, Ohio, 1939.
	500° Isothermal	O. Bauer and M. Hansen, "Metals Handbook", Cleveland, Ohio, 1939.
Cu–Ti–Zn		W. Heine and U. Zwicker, *Z. Metallkunde*, 1962, **53**, 368; E. Ence and H. Marjolin, *Trans. metall. Soc. A.I.M.E.*, 1961, **221**, 320.
Fe–Mn–Ni	Liquidus	*International Critical Tables*, 1927, **2**, 455.
Fe–Mo–Ni	1200° Isothermal	D. K. Das, S. P. Rideout and P. A. Beck, *Trans. Amer. Inst. Min. Met. Eng.*, 1952, **194**, 1071.
Fe–Nb–Si		H. J. Goldschmidt, *J. Iron & Steel Inst.*, 1960, **194**, 169.
Fe–Ni–P		G. Masing and R. Vogel, "Handbuch der Metall Physik", vol. 2, Leipsig, 1937.
Fe–Ni–Si		E. R. Greiner and E. R. Jette, *Trans. Amer. Inst. Min. Met. Eng.*, 1943, **152**, 48; S. Takede, M. Iwame and A. Sakakura, *Nippon kink. Gakk.*, 1960, **24**, 534.
Fe–Ni–W	Liquidus	G. Masing and R. Vogel, "Handbuch der Metall Physik", vol. 2, Leipsig, 1937.
Fe–O–Ti		W. Rostoker, *T.A.I.M.M.E.*, 1955, **203**, 113.
Fe–Si–Sn		R. Vogel and H. J. Jungclaus, *Arch. Eisenhüttenw.*, 1960, **31**, 243.
Fe–Sn–Zr		L. E. Tanner and D. W. Levinson, *Trans. Amer. Soc. Metals*, 1959, **52**, 166.
Hg–Mg–Mn		H. Nowotny, *Metallforsch.*, 1946, **1**, 130.

In–Pb–Sb		D. R. Geiss and E. A. Peretti, *J. less-common Metals*, 1962, **4**, 523.
In–Pb–Th		G. V. Raynor and J. Graham, *Trans. Faraday Soc.*, 1958, **54**, 161.
Li–Mg–Zn		A. F. Weinberg, D. W. Levinson and W. Rostoker, *Trans. Amer. Soc. Met.*, 1956, **48**, 855.
Mg–Mn–Th		M. E. Drits, M. V. Maltsev, E. M. Padazhnova and G. M. Bordini, *Isslednie. Splav. tsvet. Metall.*, 1960, **2**, 114.
Mg–Si–Zn	Liquidus	" Gmelin's Handbuch der anorganischen Chemie ", Mg A.4, 8th Ed., 1952.
Mg–Sn–Zn	Liquidus	" Gmelin's Handbuch der anorganischen Chemie ", Mg A.4, 8th Ed., 1952.
Mo–Ni–Si		E. W. Guard and E. A. Smith, *J. Inst. Metals*, 1959/60, **88**, 283.
Mo–Ni–Ti		R. W. Guard and E. A. Smith, *J. Inst. Metals*, 1959/60, **88**, 369.
Mo–O–Ti		P. A. Farrar, L. P. Stone and H. Margolin, *T.A.I.M.M.E.*, 1956, **206**, 595.
Mo–Ti–V		I. I. Kornilov and R. S. Polyakova, *Izv. Akad. Nauk S.S.S.R.*, 1960, **1**, 85.
Nb–W–Zr		E. M. Savitsky and A. H. Zakarhov, *Zh. neorg. Khim.*, 1962 ,**7**, 2575.
Ni–O–Ti		W. Rostowker, *T.A.I.M.M.E.*, 1955, **203**, 113.
Ni–Pd–Mn		W. Köster and M. Sallam, *Z. Metallkunde*, 1958, **49**, 240.
O–Ti–Zr		M. Hoch, R. L. Dean, C. K. Hwu and S. M. Wolosin, *Trans. metall. Soc. A.I.M.E.*, 1961, **221**, 1162.
Pb–Sb–Sn	Liquidus	F. D. Weaver, *J. Inst. Metals*, 1935, **56**, 224; V. A. Kogan and A. A. Semyonov, *Zh. fiz. Khim.*, 1963, **37**, 802.
	2 Sectionals	R. I. Jaffee and H. P. Nielsen, " Metals Handbook ", Cleveland, Ohio, 1948.
Pb–Sb–Zn	Phases in solid	*International Critical Tables*, 1927, **2**, 417.
Pb–Sn–Zn	Liquidus	R. I. Jaffee and H. P. Nielsen, " Metals Handbook ", Cleveland, Ohio, 1948.
Re–Ta–W		J. A. Brophy, M. H. Kamden and J. Wulff, *Trans. metall. Soc. A.I.M.E.*, 1961, **221**, 1137.
Si–V–Zr		M. S. Farkas, A. A. Bauer and R. F. Dickerson, *Trans. Amer. Soc. Metals*, 1959, **62**, 571.
Sn–Ti–V		W. Köster and K. Haug, *Z. Metallkunde*, 1957, **48**, 327.
Ti–U–Zr		B. W. Howlett, *J. nucl. Mater.*, 1959, **1**, 259.

QUATERNARY SYSTEMS

(No diagrams for these systems have been included)

Ag–Au–Cu–Ni	N. Parravano, *Gazz. chim. ital.*, 1914, **44**, 279.
Ag–Cd–Cu–Zr	E. Gebhardt and G. Petzow, *Z. Metallkunde*, 1960, **51**, 221.
Ag–Cu–Fe–S	E. Lüder, *Metall u. Erz.*, 1924, **21**, 329.
Ag–Cu–Mn–Zn	K. Moeller, *Z. Metallkunde*, 1943, **35**, 27.
Ag–Cu–Pb–S	W. Guertler and E. Lüder, *Metall u. Erz.*, 1924, **21**, 355; R. Schwarz and A. Romero, *Z. anorg. Chem.*, 1927, **162**, 149.
Al–Bi–Pb–Sb	G. W. Kasten, *Wiss. Veröff, Siemens-Werke*, 1940, 50.
Al–Co–Fe–Ni	G. V. Raynor and M. B. Waldron, *Proc. Roy. Soc.*, 1950, A.**202**, 420.
Al–Co–Cu–Fe	G. V. Raynor and B. J. Ward, *J. Inst. Metals*, 1957/8, **86**, 182.
Al–Cr–Mg–Zn	K. Little, H. J. Axon and W. Hume–Rothery, *J. Inst. Metals*, 1948/9, **75**, 39.
Al–Cr–Ni–Ti	L. I. Pyrakhima and L. A. Ryabsev, *Izv. Akad. Nauk S.S.S.R.*, 1957, **12**, 38.
Al–Cu–Fe–Mg	G. Phragmen, *J. Inst. Metals*, 1950, **77**, 489.
Al–Cu–Fe–Mn	G. Phragmen, *ibid.*, 1950, **77**, 489; Nabara Komatsu, *J. Inst. Met. Japan*, 1950, **14B**, 36.
Al–Cu–Fe–Ni	G. V. Raynor and B. J. Ward, *J. Inst. Metals*, 1957/8, **86**, 135.
Al–Cu–Fe–Si	A. G. C. Gwyer, H. W. L. Phillips and L. Mann, *J. Inst. Metals*, 1928, **40**, 329.
Al–Cu–Mg–Mn	G. Phragmen, *ibid.*, 1950, **77**, 489.
Al–Cu–Mg–Ni	K. E. Bingham, *ibid.*, 1926, **36**, 143.
Al–Cu–Mg–Si	*International Critical Tables*, 1927, **2**, 411; M. L. V. Gayler, *J. Inst. Metals*, 1922, **28**, 213; M. L. V. Gayler, *ibid.*, 1923, **30**, 139; A. Schrader, *Metall.*, 1949, **3**, 150; D. A. Petrov and N. D. Nagorskaya, *Zhur Obsch. Khim.*, 1949, **19**, 1994; G. Phragmen, *J. Inst. Metals*, 1950, **77**, 489; H. J. Axon, *ibid.*, 1952/3, **81**, 209.
Al–Cu–Mg–Zn	A. Saulnier, *Compt. Rend.*, 1948, **226**, 181; D. J. Strawbridge, W. Hume-Rothery and A. T. Little, *J. Inst. Metals*, 1948, **74**, 191; H. Watanabe, *Nippon kink. Gakk.*, 1959, **23**, 596.
Al–Cu–Mn–Si	G. Phragmen, *J. Inst. Metals*, 1950, **77**, 489.
Al–Fe–Mg–Mn	G. Phragmen, *J. Inst. Metals*, 1950, **77**, 489.
Al–Fe–Mg–Si	H. W. L. Phillips, *J. Inst. Metals*, 1946, **72**, 168; A. Schrader, *Metall.*, 1949, **3**, 150; G. Phragmen, *J. Inst. Metals*, 1950, **77**, 489.
Al–Fe–Mn–Si	H. W. L. Phillips and P. C. Varley, *J. Inst. Metals*, 1943, **69**, 324; G. Phragmen, *ibid.*, 1950, **77**, 489.
Al–Mg–Mn–Si	G. Phragmen, *ibid.*, 1950, **77**, 489.
Al–Mg–Mn–Zn	A. T. Little, G. V. Raynor and W. Hume-Rothery, *J. Inst. Metals*, 1943, **69**, 423; *ibid.*, 1943, **69**, 467.
Al–Mg–Si–Zn	W. Sander and K. L. Meissner, *Z. Metallkunde*, 1923, **15**, 180; *ibid.*, 1924, **16**, 12; F. Bollenrath and H. Gröber, *Metallforsch.*, 1946, **1**, 116.
Bi–Cd–Pb–Sn	N. Parravano and G. Sirovich, *Gazz. chim. ital.*, 1911, **41**, 654; *ibid.*, 1912, **42**, 630.
C–Cr–Fe–Ni	E. C. Bain and R. H. Aborn, "Metals Handbook", Cleveland, Ohio, 1939.
C–Cr–Fe–Si	T. Murakami and K. Yokoyama, *Tech. Rep. Tôhoku. Imp. Univ.*, 1931, **10**, 245; E. Valenta and F. Poboril, *Chim. et Ind.*, 1933, 633.
C–Cu–Fe–Mn	F. Ostermann, *Z. Metallkunde*, 1925, **17**, 278.
C–Fe–Mn–S	T. Satô, *Tech. Rep. Tôhoku. Imp. Univ.*, 1935, **11**, 234; H. Wentrup, *Carnegie Schol. Mem.: Iron & Steel Inst.*, 1935, **24**, 112.
C–Fe–Mn–Si	O. v. Keil and F. Kotyza, *Arch. Eisenhüttenwesen*, 1930, **4**, 295.
Cr–Co–Mo–Ni	S. P. Rideout and P. A. Beck, *U.S. Nat. Advis. Cttee. Aeronautics Rep.*, 1953 (1122).
Cr–Fe–Mn–N	H. Krainer and O. Mirt, *ibid.*, 1942, **15**, 467.
Cr–Fe–Mo–Ni	C. J. Bechtold and H. A. Vacher, *J. Res. Nat. Bur. Stand.*, 1957, **58**, 7.

Cu–Fe–Mn–Ni   N. Parravano, *Gazz. chim. ital.*, 1912, **42**, 589.
Cu–Fe–Mn–Zn   G. Edmunds, *Trans. Am. Inst. Min. & Mech. Eng.*, 1944, **156**, 263.

Fe–Mn–O–Si   C. Benedicks and H. Löfquist, " Non-Metallic Inclusions in Iron
             and Steel ", London, 1930; W. R. Maddocks, *Carnegie Schol.
             Mem. : Iron & Steel Inst.*, 1935, **24**, 51.

Ni–Pb–Sb–Sn   C. L. Ackermann, *Z. Metallkunde*, 1929, **21**, 339.

QUINARY SYSTEMS

Al alloys with Cu, Fe, Mg, Mn, Si   G. Phragmen, *J. Inst. Metals*, 1950, **77**, 489.
Al–Cr–Fe–Ni–Ti   A. Taylor, *T.A.I.M.M.E.*, 1957, **209**, 72.
Al–Cr–Ni–Ti–W   I. I. Kornilov, L. I. Pryskhina and O. V. Ozhimkova, *Izvest. Akad.
             Nauk. S.S.S.R.*, 1956, 885.

GAS–METAL SYSTEMS

THE SOLUTION OF GASES IN METALS

SOLUTIONS of gases in metals may be divided into:

(a) equilibrium solutions of the diatomic gases hydrogen, oxygen and nitrogen.

(b) non-equilibrium solutions of the inert gases helium, argon, krypton and xenon.

The diatomic gases in solution can establish equilibrium with the gas phase via a chemisorbed monolayer of atoms on the free surface and the term solubility denotes the concentration of dissolved gas in equilibrium with a specified pressure of the gas. Usually, it is found that when no second phase is present, the concentration is proportional to the square root of the pressure, indicating that these gases dissociate into atoms during the solution process. Assuming the activity of the dissolved gas is equal to the mole fraction, $[S]$, the equilibrium constant is given by

$$K = [S]^2/p \qquad \cdots \cdots \cdots \quad (1)$$

For many systems a plot of log. solubility at constant pressure against the reciprocal of the absolute temperature is linear, so that the heat of solution, ΔH can be calculated using equation (1) and the simple form of the Van't Hoff equation

$$2\ln S_1/S_2 = \ln K_1/K_2 = \frac{-\Delta H}{R}\frac{(T_2 - T_1)}{T_1 T_2} \qquad \cdots \cdots \quad (2)$$

where K_1, K_2 are the equilibrium constants at temperatures T_1, T_2 and S_1, S_2 are the corresponding solubilities at some constant reference pressure.

There are examples of both endothermic solution where solubility rises with temperature and exothermic solution where solubility decreases with temperature.

In systems where compounds are formed, an alternative convention is sometimes adopted in which the data given is the solubility of the compound, defined as the concentration of dissolved gas in equilibrium with the free compound (i.e. with a pressure in the gas phase equal to the dissociation pressure of the compound). Since formation of the compound may be a strongly exothermic process, the solution of the gas may be endothermic with respect to the compound even though it is exothermic with respect to the gas phase. The variation with temperature can be represented by an equation of the form,

$$\log_{10}[\%X]/(MX) = -A/T + B$$

where (MX) represents the activity of the compound.

The inert gases do not dissolve in true metals from a neutral gas phase since their atoms are not chemisorbed on metallic surfaces. Solutions are, however, produced by:

(a) fission of the solvent (U $\longrightarrow$ Kr, Xe);

(b) injections of ions (e.g. He$^+$) accelerated by particle accelerators into massive samples;

(c) injections of ions into thin films by electrical discharge.

Inert gas solutes have some of the characteristics of other gas solutes, e.g. diffusion coefficients can be assigned to them, but since equilibrium with the gas phase cannot be established the term solubility cannot have the normal quantitative significance.

References 1–7 give general reviews on the subject of gas–metal equilibria. References 8–12 deal specifically with hydrogen in metals. Reference 13 gives a theoretical analysis of inert gas solutions in metals.

SOLUBILITY OF HYDROGEN IN METALS

Simple physical solutions

Metals in these systems dissolve small quantities of hydrogen endothermically. An exception is manganese in which the solution is exothermic in the low-temperature α-phase. The metals do not form hydrides under conditions of metallurgical interest. Data are given in Table 1 for atmospheric pressure and values for other

TABLE I.　SOLUBILITY OF HYDROGEN IN METALS WHICH DISSOLVE LOW CONCENTRATIONS OF THE GAS

Metal			Solubility									ΔH	References
Ag (solid)	Temp. (°C) ...	S760 ...	400 0·056	500 0·11	600 0·18	700 0·23	800 0·33	900 0·43	960·5+ 0·52			$\Delta H = 12{,}000$	14
Al (solid)	Temp. (°C) ...	S760 ...	350 —	400 0·005 / 0·0012	500 0·013 / 0·0028	600 0·026 / 0·011	660+ 0·036 / 0·030	0·050				$\Delta H = 19{,}000$ / $\Delta H = 27{,}800$	15 / 16
(liquid)	Temp. (°C) ...	S760 ...	660+ 0·69 / 0·43	700 0·92 / 0·63 / 0·90	750 1·23	800 1·67 / 1·23 / 1·75	900 2·18 / 2·75	1000 3·51 / 4·15				$\Delta H = 25{,}000$ / $\Delta H = 28{,}210$ / $\Delta H = 23{,}300$	15 / 16 / 17
Al–Cu (liquid)	Temp. (°C) S760 at 2% Cu	4% Cu ...	8% Cu ...	16% Cu ...	32% Cu ...	700: 0·75 0·65 0·50 0·40 0·35	800: 1·35 1·15 0·95 0·80 0·65	900: 2·45 2·25 1·85 1·35 1·15	1000: 3·80 3·45 2·95 2·30 1·80			$\Delta H = 27{,}000$ / 27,900 / 28,800 / 28,800 / 27,000	17 17 17 17 17
Al–Si (liquid)	Temp. (°C) S760 at 2% Si	4% Si ...	8% Si ...	16% Si ...		700: 0·75 0·70 0·60 0·50	800: 1·50 1·30 1·25 1·23	900: 2·50 2·35 2·25 2·15	1000: 3·90 3·75 3·60 2·37			$\Delta H = 25{,}600$ / — / 27,900 / 28,800	17 17 17 17
Co (solid)	Temp. (°C) ...	S760 ...	600 0·9	700 1·22	800 1·85	900 2·51	1000 3·30	1100 4·31	1200 5·40			$\Delta H = 16{,}500$	18
(liquid)	Temp. (°C) ...	S760 ...	1600 23·2										19
Co–Fe	See Fe–Co												
Co–Ni	See Ni–Co												
Cr (solid)	Temp. (°C) ...	S760 ...	400 0·2	500 0·4 / 0·3	600 0·6 / 0·4	700 0·8 / 0·6	800 1·0 / 1·0	900 1·8 / 1·7	1000 3·0 / 2·6	1100 4·3 / 3·7	1200 5·4		20 / 21
Cr–Fe	See Fe–Cr												
Cu (solid)	Temp. (°C) ...	S760 ...	300 0·01	400 0·05 / 0·06	500 0·11 / 0·16	600 0·21 / 0·31 / 0·11	700 0·49 / 0·27	800 0·72 / 0·53	900 1·08 / 0·89	1000 1·57 / 1·34	1083+ 2·1 / 1·9		22 23 24

(liquid) — continued

Temp. (°C)	1083+	1100	1150	1200	1250	1300	1350	1400	1500	Ref
S_{760}	6·0	6·3	7·3	8·3	9·3	10·3	11·1	12·0	13·3	23
S_{760}	5·1	5·4	6·3	7·2	8·3	9·2	10·4	11·8	13·3	24
S_{760}	5·4	5·7	—	7·3	—	9·4	—	—	—	25

Cu–Ag (liquid) — Data for 1225°C (Ref 5)

Wt-% Ag	0	10	20	30	50
S_{760}	8·6	7·9	7·0	5·9	4·3

Cu–Al (solid) (Ref 24)

Temp. (°C)	700	800	900	1000	1050
S_{760} at 1·43% Al	—	—	0·70	1·15	1·40
,, 3·3% Al	—	0·35	0·70	1·05	1·25
,, 5·77% Al	0·10	0·10	0·30	0·60	—
,, 6·84% Al	0·10	0·15	0·35	0·65	—
,, 8·1% Al		0·15	0·35	0·75	—

Cu–Al (liquid) (Ref 24)

Temp. (°C)	1100	1150	1200
S_{760} at 1·45% Al	5·1	5·8	6·6
,, 3·3% Al	4·4	5·1	5·7
,, 5·77% Al	3·3	3·9	4·5
,, 6·84% Al	3·0	3·5	4·1
,, 8·1% Al	2·7	3·2	3·5

Data for higher Al concentrations at 1225°C (Ref 5):

Wt-% Al	0	10	20
S_{760}	8·6	3·6	2·5

Cu–Au (liquid) — Data for 1225°C (Ref 5)

Wt-% Au	0	30	50
S_{760}	8·6	5·6	3·4

Cu–Ni (liquid) — Data for 1225°C (Ref 5)

Wt-% Ni	0	10	20
S_{760}	8·6	13·0	17·0

Cu–Pt (liquid) — Data for 1225°C (Ref 5)

Wt-% Pt	0	10	20
S_{760}	8·6	9·6	9·7

Cu–Sn (liquid) (Ref 25)

Temp. (°C)	1000	1100	1200	1300
S_{760} at 5·9% Sn	—	4·80	6·28	7·81
,, 11·5% Sn	3·09	4·11	5·35	6·85
,, 21·7% Sn	2·11	2·97	3·94	5·10
,, 40·2% Sn	0·53	0·94	1·50	—
,, 54·8% Sn	0·50	0·76	1·15	1·61

Alternative data for 1225°C (Ref 5):

Wt-% Sn	0	10	20	30	40	50	60
S_{760}	8·6	7·1	5·3	3·8	2·3	1·9	1·4

Cu–Zn (solid, α) (Ref 26)

Temp. (°C)	500	600	700	800	875*
S_{760} at 33% Zn	0·010	0·023	0·044	0·076	0·35

(* α + β region)

S_{760} = Solubility at atmospheric pressure expressed as cc NTP/100 g of metal.

ΔH = Heat of solution calculated from solubility data in cal/g mol. of hydrogen.

+ = Indicates melting points.

TABLE I. SOLUBILITY OF HYDROGEN IN METALS WHICH DISSOLVE LOW CONCENTRATIONS OF THE GAS—continued

Fe (solid, α)

Temp. (°C)	200	300	400	500	600	700	800	900	ΔH	References
S_{760}	0·05	0·19	0·46	0·61	1·00	1·45	—	—	$\Delta H = 11{,}600$	27
	—	—	—	0·6	—	1·8	—	3·0		28
	—	0·16	0·38	0·72	1·18	1·75	2·40	3·14	$\Delta H = 13{,}300$	29
	—	0·1	0·2	0·6	1·2	1·7	2·4	3·0		21

Fe (solid, γ)

Temp. (°C)	900	1000	1100	1200	1250	1350	1400	References
S_{760}	4·7	5·4	7·0	8·2	8·8	10·1	10·5	28
	4·7	—	6·4	7·4	8·4	—	9·3	21
	—	—	—	—	—	—	9·0	30

Fe (solid, δ)

Temp. (°C)	1400	1450	1535+	References
S_{760}	10·1	11·0	14·0	28
	12·0	—	—	30
	6·1	6·4	—	21

Fe (liquid)

Temp. (°C)	1535+	1550	1560	1600	1650	1685	ΔH	References
S_{760}	25·0	26·0	—	28·4	31·0	—		28
	22·5	23·0	—	29·8	26·0	—		31
	—	29·2	—	—	32·5	—		32
	—	—	—	—	—	—		30
	—	—	23·6	—	—	—		19
	—	—	—	—	—	27·2	$\Delta H = 16{,}000$	33

Mild Steel

Data for a typical steel containing 0·39% C, 0·70% Mn, 0·13% Si:

Temp. (°C)	400	500	600	References
*S_{760}	0·48	0·80	1·35	29

*Calculated from data determined at 1 mm Hg pressure

Fe–Nb (liquid)

Temp. (°C)	1560	1685	ΔH	References
*S_{760} at 5·01% Cb	24·6	28·3	$\Delta H = 16{,}000$	33
* ,, 8·97% Cb	29·2	31·7	$\Delta H = 9{,}900$	33
* ,, 15·12% Cb	41·7	43·8	$\Delta H = 5{,}600$	33

*Calculated from data determined at 22 mm Hg pressure

Fe–Co (liquid)

Data for 1600°C:

Wt-% Co	20	40	60	80	References
S_{760}	23·8	20·7	18·2	21·5	19

Fe–Cr (solid)

Temp. (°C)	400	600	700	850	900	1000	1100	1200	References
S_{760} at 4% Cr	0·2	0·9	1·4	2·2	4·4	5·3	6·3	7·4	21
,, 10% Cr	0·2	0·7	1·2	5·0	6·0	7·1	8·3	9·3	21
,, 20% Cr	0·2	0·5	1·8	3·7	4·3	4·9	5·6	6·6	21
,, 44% Cr	0·2	0·5	1·2	3·5	4·3	5·4	6·2	6·9	21
,, 77% Cr	0·2	0·5	0·8	1·5	2·0	3·1	4·4	5·9	21

Data for 1400°C:

Wt-% Cr	9·4	18·8	28·5	38·3	48·2	References
S_{760}	9·0	9·0	8·0	8·0	7·0	30

Fe-Cr (liquid)

Data for 1600°C:

Wt-% Cr	9.4	18.8	28.5	38.3	48.2	Ref.
S_{760}	25.0	25.0	25.0	24.0	23.0	30

Fe-Ni (solid)

Temp. (°C)	300	400	500	600	700	800	900	1000	1100	1200	Ref.
S_{760} at 3.3% Ni	—	0.3	0.7	1.3	2.0	3.9	5.0	6.2	7.3	8.4	21
,, 5.5% Ni	0.2	0.4	0.8	1.4	2.1	4.1	5.2	6.3	7.3	8.4	21
,, 11.6% Ni	0.4	0.6	1.3	2.1	3.2	4.3	5.3	6.4	7.5	8.5	21
,, 21.0% Ni	0.5	0.7	1.4	2.2	3.3	4.3	5.4	6.5	7.6	8.8	29
* ,, 28.0% Ni	0.6	0.93	1.51	2.19	—	—	—	—	—	—	21
,, 32.1% Ni	0.6	0.9	1.5	2.4	3.3	4.4	5.5	6.7	7.8	9.0	21
,, 53.2% Ni	0.8	1.2	1.8	2.7	3.7	4.7	5.8	6.9	8.1	9.3	21
,, 62.4% Ni	1.3	1.7	2.4	3.3	4.1	5.0	6.1	7.3	8.4	9.6	21
,, 72.4% Ni	1.6	1.8	2.6	3.4	4.3	5.2	6.3	7.4	8.7	9.9	21
,, 84.8% Ni	1.8	2.2	2.7	3.7	4.8	6.3	7.6	9.2	10.7	12.4	21

Data for 1400°C:

Wt-% Ni	10.3	20.6	30.8	60.8	70.8	80.6	90.3	Ref.
,, S_{760}	14.0	15.0	15.0	15.0	16.0	16.0	17.0	30

* Calculated from data determined at 1 mm Hg pressure

Fe-Ni (liquid)

Data for 1600°C:

Wt-% Ni	10.3	20.6	30.8	40.9	60.8	70.8	80.6	90.3	Ref.
S_{760}	29.0	30.0	31.0	32.0	34.0	35.0	36.0	39.0	30

Alternative data for 1600°C:

Wt-% Ni	20	40	60	80	90	Ref.
S_{760}	30.6	31.3	33.6	39.5	42.3	19

Fe-Si (liquid)

Temp. (°C)	1350	1400	1500	1550	1650	Ref.
S_{760} at 1.78% Si	11.5	12.5	25.5	27.5	31.6	32
,, 11.0% Si	7.0	7.6	14.4	15.4	17.4	32
,, 21.7% Si	—	6.0	8.9	9.5	10.7	32
,, 31.5% Si	—	7.0	6.5	6.8	7.3	32
,, 39.1% Si	9.3	9.9	8.5	9.1	10.3	32
,, 45.7% Si	12.7	13.1	11.1	11.8	13.1	32
,, 51.5% Si	—	—	14.0	14.4	15.3	32
,, 63.7% Si	20.9	21.4	22.4	22.9	23.8	32

Fe-Ti (liquid)

Temp. (°C)	1560	1685		Ref.
S_{760} at 0.18% Ti	24.3	26.7	$\Delta H = 10{,}900$	33
,, 0.45% Ti	25.6	28.2	$\Delta H = 10{,}900$	33
,, 0.70% Ti	28.2	30.6	$\Delta H = 8{,}800$	33
,, 2.62% Ti	—	45.8	$\Delta H = $ —	33
,, 3.14% Ti	50.0	52.9	$\Delta H = 5{,}600$	33

Fe-Cr-Ni (solid)

Temp. (°C)	400	500	600	700	800	900	1000	1100	1200	Ref.
S_{760} at 17.5% Cr + 9.0% Ni	0.2	0.8	1.8	2.4	3.4	4.9	6.4	7.7	8.8	21
,, 4.6% ,, 10.0% Ni	0.6	1.0	1.7	2.8	3.6	4.4	5.5	6.6	8.1	21
* ,, 18.0% ,, 8.0% Ni	3.24	3.95	4.67	—	—	—	—	—	—	29

* Calculated from data determined at 1 mm Hg pressure
Data for certain other compositions are given in Reference 21

S_{760} = Solubility at atmospheric pressure expressed at cc NTP/100 g of metal.
ΔH = Heat of solution calculated from solubility data in cal/g mol. of hydrogen.
† = Indicates melting points.

TABLE I.   SOLUBILITY OF HYDROGEN IN METALS WHICH DISSOLVE LOW CONCENTRATIONS OF THE GAS—*continued*

Mg (solid) — References 34, 35, 37

Temp. (°C)	640	650+	700	725	800	900
S760	30·7	~20	~26	—	~27	~30
	—	~15	—	—	~23–25	—

Mg (liquid) — References 34, 35, 36, 37

Temp. (°C)	650+	675	700	725	760	775
S760	~25	46·5	~26	60·1	~41	63·1

Data not reliable

Mg–Al (liquid) — Reference 35

Data for an unspecified temperature:

Wt-% Al	20	40	60	80
S760	16·0	11·0	8·0	5·0

Mn (solid, α) — References 38, 39

Temp. (°C)	25	100	200	300	400	500	600
S760	43·1	24·1	17·3	11·1	8·9	8·9	9·1
	21·6	19·9	17·2	14·5	12·4	11·4	11·4

α ⟶ β transition — References 38, 38, 39, 39

Temp. (°C)	600	625	650	675	700	725	750	775	800
(1) S760 rising temp.	9·0	9·1	9·2	9·4	9·8	10·2	11·1	25·4	27·2
falling temp.	9·0	9·1	9·2	9·4	9·8	10·2	11·1	25·4	27·2
(2) S760 rising temp.	11·4	11·4	11·6	12·0	25·4	26·0	26·5	26·8	27·2
falling temp.	11·4	11·4	14·8	26·5	27·8	28·0	28·2	28·4	28·6

(solid, β & γ) — References 38, 39

Temp. (°C)	800	850	900	950	1000	1050	1100	1125
S760	27·2	27·8	28·6	29·6	31·0	33·2	39·0	41·7
	28·6	29·3	30·1	31·4	32·8	34·2	40·0	42·2

γ ⟶ δ transition — References 39, 39

Temp. (°C)	1125	1130	1135	1140	1145	1150
S760 rising temp.	42·2	42·3	42·5	42·5	41·5	41·1
falling temp.	42·2	42·3	41·6	41·1	41·1	41·1

(solid, δ) — References 38, 39

Temp. (°C)	1175	1200	1225	1243+
S760	39·6	41·0	43·0	51·8
	41·7	42·8	44·4	46·6

(liquid) — References 38, 39

Temp. (°C)	1250	1275	1300
S760	51·8	60·0	61·0
	50·0	58·3	60·2

Mo (solid) — Reference 20

Temp. (°C)	500	600	700	800	900	1000	1100	1200
S760	0·8	1·3	1·7	2·2	1·8	1·2	0·8	0·5

Ni (solid)

Temp. (°C)	300	400	500	600	700	800	900	Ref.
S_{760}	2·0	2·5	3·3	4·3	5·6	7·0	8·5	21
	2·4	3·2	4·2	5·4	6·7	8·0	9·5	23
	2·49	3·68	4·88	6·07	7·24	8·35	9·40	29

$\Delta H = 5{,}900$

Temp. (°C)	1000	1100	1200	1300	1400	1453+	Ref.
S_{760}	10·0	11·5	13·0	15·9	17·5	18·8	21
	11·1	12·7	14·3	—	18·8	18·3*	23
							30

* By extrapolation

Ni (liquid)

Temp. (°C)	1453+	1500	1570	1600	Ref.
S_{760}	39·2*	—	—	42·5	19
	—	40·9	—	43·4	23
	—	38·5	38·8	43·5	30
	—	—	—	40·3	31

* By extrapolation

Ni-Co (liquid) — Data for 1600°C:

Wt-% Co	20	40	60	80	Ref.
S_{760}	24·9	19·8	21·2	22·7	19

Ni-Fe — See Fe-Ni

Pb (liquid)

Temp. (°C)	420	500	600	700	800	900	Ref.
S_{760}	0·0	0·11	0·25	0·45	0·80	1·25	40
	0·0	0·0	—	—	—	—	41
	—	<0·01	—	—	—	—	42

$\Delta H = 22{,}300$

Pb-Ca (liquid) — Hydrogen is insoluble in lead with 0·16 or 0·24% Ca at 420°C — Ref. 41

Pb-Mg (liquid) — Data for 500°C:

Wt-% Mg	1·0	2·1	3·1	5·5	6·5	7·6	7·8	8·5	9·0	17·5	Ref.
S_{760}	0·18	0·22	0·45	0·87	1·7	1·5	1·8	2·4	1·9	4·8	41

Pt (solid)

Temp. (°C)	400	800	1000	1100	1200	1300	Ref.
S_{760}	0·07	0·10	0·19	0·34	0·54	0·80	43

Si (solid) — S_{760} at 1200°C = 0·0016 — Ref. 44

Sn (liquid)

Temp. (°C)	1000	1100	1200	1300	Ref.
*S_{760}	0·04	0·09	0·21	0·36	25

* Averages. Results of replicate determinations show wide scatter

Zn (liquid) — At 516°C the solubility of hydrogen is <0·002 cc NTP/100g — Ref. 42

S_{760} = Solubility at atmospheric pressure expressed as cc NTP/100 g of metal.
ΔH = Heat of solution calculated from solubility data in cal/g mol. of hydrogen.
+ = Indicates melting points.

pressures can be obtained by assuming that the equilibrium concentration is proportional to the square root of the pressure. Since determinations can be subject to considerable experimental error, confirmatory data from two or more sources are given where possible. All data have been converted to units of cc NTP/100 g for easy comparison.

Systems in which salt-like hydrides form

Metals in these systems form distinctive hydrides (*e.g.* with electrovalent structures). The solubility of hydrogen is usually high and terminal solid solutions of hydrogen in the metal are in equilibrium with the hydride under its dissociation pressure at any chosen temperature, giving pressure-concentration curves of the form shown in *Figure 1*.

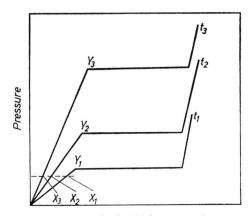

Hydrogen (or hydride) concentration

Figure 1

Data may be determined or presented as:

(*a*) The solubility of hydrogen at various temperatures at standard reference pressures (*e.g.* 760 mm mercury pressure), giving concentrations at points such as X_1, X_2, etc., in *Figure 1*.

(*b*) The terminal solubility of hydrogen at the dissociation pressures of the hydride at selected temperatures, *i.e.* concentrations at Y_1, Y_2, etc., in *Figure 1*. Sometimes this data is given in the form of the solubility of the hydride, *i.e.* the values given include the stoichiometric proportion of metal corresponding to the dissolved hydrogen.

Up to the values corresponding to the dissociation pressures the quantity of dissolved hydrogen is usually proportional to the square root of the pressure in the gas phase and solubilities at lower pressures may be calculated on this basis.

The Barium–Hydrogen System

TABLE 2. SOLUBILITY OF HYDRIDE (BaH_2) AT ITS DISSOCIATION PRESSURE
(PETERSON and INDIG[45])

Temp. (°C)	400	500	600	700
Solubility (mole fraction of BaH_2)	0·11	0·16	0·27	0·39

The Calcium–Hydrogen System

TABLE 3. SOLUBILITY OF HYDRIDE (CaH_2) AT ITS DISSOCIATION PRESSURE (TREADWELL and STECHER[46])

Temp. (°C)	780	800	830	860
Dissociation Pressure (mm Hg)	14·0	27·2	47·0	81·0
Solubility (%CaH₂)	18·0	20·5	22·0	23·0

Other references: JOHNSON et al.[47] (confirmatory data); PETERSON and FATTORE [48] (phase diagram Ca–CaH₂)

The Magnesium–Hydrogen System

TABLE 4. SOLUBILITY OF HYDROGEN AT DISSOCIATION PRESSURE OF MgH_2 (STAMPFER, HOLLEY and SHUTTLE[49])

Temp. (°C)	440	470	510	560
Dissociation Pressure (atm.)	38·5	64·7	117·1	233·0
*Solubility (mole fraction of H)	0·020	0·031	0·034	0·093

*Estimated error ±0·007.

Other references: KENNELEY et al.[50] (hydride dissociation pressures. See also Table 1 for S_{760})

The Sodium–Hydrogen System

TABLE 5. SOLUBILITY OF HYDROGEN IN MOLTEN SODIUM AT THE DISSOCIATION PRESSURE OF NaH (ADDISON, PULHAM and ROY,[91] WILLIAMS, GRAND and MILLER[92])

Temp. (°C)	250	300	315	330	350	375	400	425
Solubility (wt-%H)	0·00042	0·0022	0·0052	0·0104	0·013	0·025	0·062	0·12

The Uranium–Hydrogen System

TABLE 6. SOLUBILITIES OF HYDROGEN: (a) S_{760} AT 1 ATM PRESSURE; AND (b) S_T AT HYDRIDE DISSOCIATION PRESSURES P EXPRESSED IN P.P.M. (MALLET and TRZECIAK[51])

Phase		Solubility				
α	Temp. (°C)	100	200	300	400	432
	S_{760} (p.p.m.)	—	—	—	—	1·6
	P (atm)	2 × 10⁻⁶	6 × 10⁻⁴	0·03	0·5	1·0
	S_T (p.p.m.)	6 × 10⁻⁴	0·02	0·2	1·1	1·6
	Temp. (°C)	450	500	550	600	662
	S_{760} (p.p.m.)	—	1·8	—	2·0	2·2
	P (atm)	1·5	3·7	8·5	17·9	39·7
	S_T (p.p.m.)	2·0	3·5	5·6	8·6	13·5
β	Temp. (°C)	662	700	725	750	769
	S_{760} (p.p.m.)	7·8	8·5	9·0	—	9·7
	P (atm)	39·7	61·6	81·2	104	129
	S_T (p.p.m.)	49	68	81	97	111
γ	Temp. (°C)	769	800	850	900	950
	S_{760} (p.p.m.)	14·7	15·0	—	15·6	—
	P (atm)	129	213	263	397	575
	S_T (p.p.m.)	168	195	249	312	385
	Temp. (°C)	1000	1050	1100	1129	
	S_{760} (p.p.m.)	16·2	—	16·7	16·9	
	P (atm)	793	1100	1470	1740	
	S_T (p.p.m.)	457	545	643	702	
Liquid	Temp. (°C)	1129	1200	1300	1400	
	S_{760} (p.p.m.)	28·1	29·3	31·1	32·7	
	P (atm)	1740	2450	3880	5880	
	S_T (p.p.m.)	1170	1450	1940	2520	

Other reference: MATTROW [52] (data for powdered uranium at low pressure)

Systems in which metallic hydrides are formed

Metals in these systems can absorb large atomic fractions of hydrogen and the absorption of the gas is exothermic.   The systems include compositions corresponding to hydrides of simple stoichiometric ratios at which, however, the metallic character of the material persists.   There are marked lattice changes as hydrogen is taken up.   An entirely satisfactory classification has yet to be evolved.   Some systems are entirely single phase (*e.g.* Nb–H) and transition from metal to hydride occurs continuously as hydrogen is taken up.   In others (*e.g.* Th–H), isotherms have a similar form to those found for systems in which saline hydrides are formed. In certain other systems (*e.g.* Ti–H) there are two-phase fields (in which pressure is invariant with composition) due to hydrogen-stabilization of high-temperature allotropic forms of the metal.

Isothermal pressure-concentration curves for a number of systems are given in *Figures 2–12.*   Data for hafnium are available only for atmospheric pressure and are tabulated.

The Cerium–Hydrogen System

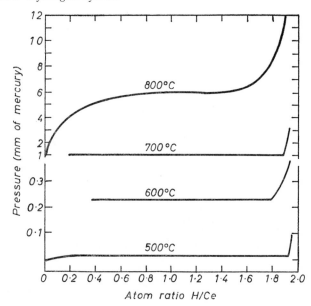

Figure 2.   Cerium–Hydrogen (MULFORD and HOLLEY [53])
See also: SIEVERTS and co-workers [54–56] (alternative data)

The Niobium–Hydrogen System

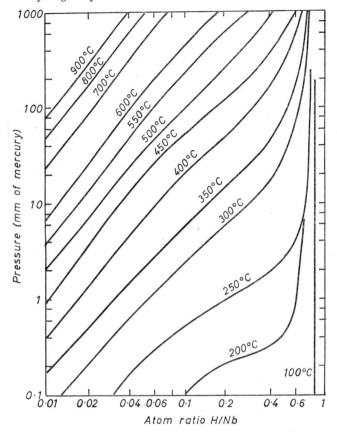

Figure 3. Niobium–Hydrogen (ALBRECHT, GOODE and MALLET [59], [60])
See also: KOMJATHY [61] (supporting data)

The Neodymium–Hydrogen System

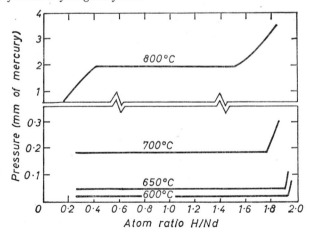

Figure 4. Neodymium–Hydrogen (MULFORD and HOLLEY [53])
See also: SIEVERTS and ROELL [62] (earlier data)

The Palladium–Hydrogen System

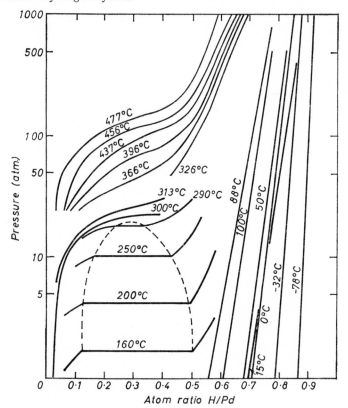

Figure 5. *Palladium–Hydrogen* (LEVINE *and* WEAL,[63] GILLESPIE and co-workers [64], [65] PERMINOV *et al.*[66])

See also: EVERETT and NORDON [67] (hysteresis); FLANAGAN [68] (absorption of deuterium); NAKHUTIN and SUTYAGINA [69] (low temperature data); MITACEK and ASTON [70] (thermodynamic properties); CARSON *et al.*[71] (H in Pd–Pt alloys); KARPOVA and TVERDOVSKY [72] (H in Pd–Cu alloys)

The Praseodymium–Hydrogen System

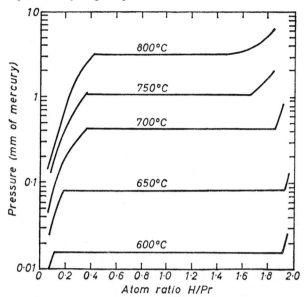

Figure 6. Praseodymium–Hydrogen (MULFORD and HOLLEY [53])
See also: SIEVERTS and ROELL [62] (earlier data)

The Plutonium–Hydrogen System

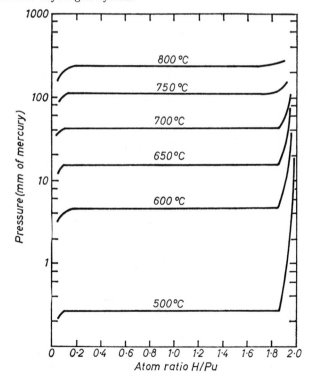

Figure 7. Plutonium–Hydrogen (MULFORD and STURDY [73])

The Tantalum–Hydrogen System

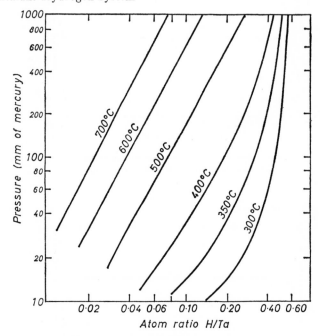

Figure 8.   Tantalum–Hydrogen (MALLET and KOEHL [74])
See also: KOFSTAD *et al.*[75] (data for temperatures to 400°C)

The Thorium–Hydrogen System

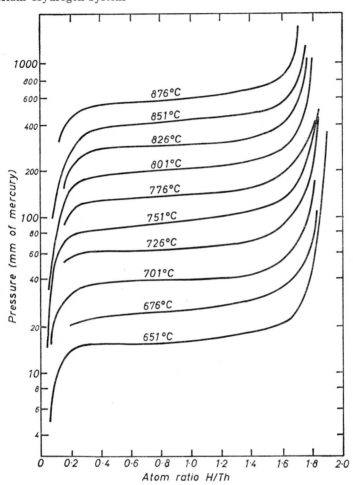

Figure 9. Thorium–Hydrogen (MALLET and CAMPBELL [76])
See also: PETERSON and WESTLAKE [77] (supporting data)

The Titanium–Hydrogen System

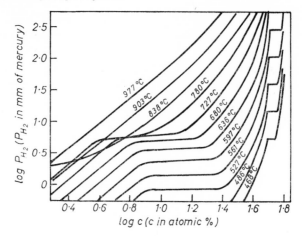

Figure 10.   Titanium–Hydrogen (McQuillan [78])
See also: Lenning *et al.*[79] (low temperature data); McQuillan [80] (Ti–Cu and Ti–Fe alloys)

The Vanadium–Hydrogen System

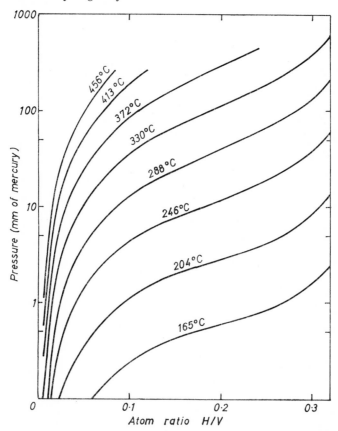

Figure 11.   Vanadium–Hydrogen (calculated from data given by Kofstad and Wallace [81])

The Zirconium–Hydrogen System

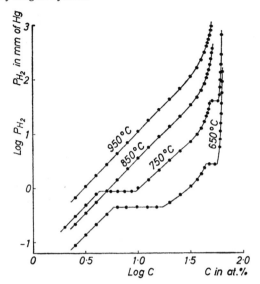

Figure 12. Zirconium–Hydrogen (Private communication from McQuillan *based on reference* 82)

See also: Motz [83] (a review); Schwartz and Mallet [84]; Gulbransen and Andrew [85]; Edwards *et al.* [86]; Mallet and Albrecht [87]; Espagno *et al.* [88]; Libowitz [89]; La Grange *et al.* [90] (references 84–90 give confirmatory data)

The Hafnium–Hydrogen System

TABLE 7.   SOLUBILITY OF HYDROGEN AT 760 MM PRESSURE (ESPAGNO, AZOU and BASTIEN [57,58])

Temp. (°C)	100	300	500	700	900	950	1000	1050	1100
Atom ratio H/Hf	1·80	1·78	1·60	1·38	0·88	0·40	0·09	0·06	0·05

SOLUBILITY OF NITROGEN IN METALS

Dilute solutions

Certain metals, notably iron, dissolve low concentrations of nitrogen. In the solid metal, the solute atoms are assumed to be accommodated interstitially but only a small proportion of the potential sites are occupied. The solubility may be enhanced if other sites are created by introducing lattice defects into the metal (see *e.g.* references 93 and 94). If a metal forms a nitride the solubility of nitrogen is terminated at a value corresponding to equilibrium with the gas at the nitride dissociation pressure.

Solubilities are usually determined by measuring the uptake of nitrogen from the gas phase and presented as concentrations in equilibrium with the gas at atmospheric pressure (S_{760}). Table 8 gives the available data in this form. Many more determinations have been made for iron and iron alloys than for other metals because of the important influence of nitrogen on the properties of steels.

TABLE 8. SOLUBILITY OF NITROGEN IN METALS WHICH DISSOLVE LOW CONCENTRATIONS OF THE GAS

Fe (solid α)

				References
Temp. (°C)	700	800	900	95
S_{760} (wt-%)	0.5×10^{-3}	2.3×10^{-3}	2.5×10^{-3}	96
	1.5×10^{-3}		3.3×10^{-3}	

Fe (solid γ)

							References
Temp. (°C)	900	1000	1100	1200	1300	1400	97
S_{760} (wt-%)	—	—	0·0242	0·0232	0·0221	0·0210	98
	—	0·025	0·024	0·023	0·022		96
	0·0268	0·0250	0·0236	0·0224	0·0214	0·0206	119
	0·0291	0·0260	—	0·0221	—	—	

Fe (solid δ)

					References
Temp. (°C)	1400	1450	1500	1535+	97
S_{760} (wt-%)	0·0101	0·0111	0·0121	0·0129	

Fe (liquid)

There have been more than 20 determinations of the solubility of nitrogen in liquid iron at temperatures near 1600°C. The temperature coefficient is small. Results given by different authors are in fair agreement and are given below:

Data for 1600°C:

									References
Reference	99	97	100	101	102	103	104	105	19, 97
S_{760} (wt-%)	0·040	0·031	0·043	0·041	0·038	0·046	0·040	0·052	99–117
Temp. coeff. (wt-%/°C × 10⁵)	1·5	—	1·0 to 2·0	—	—	3·4	—	0·2	
Reference	106	107	108	109	110	111	112		
S_{760} (wt-%)	0·039	0·044	0·046	0·044	0·042	0·044	0·044		
Temp. coeff. (wt-%/°C × 10⁵)	1·4	—	—	—	1·5	—	0·77		
Reference	114	115	116	117			113		
S_{760} (wt-%)	0·040	0·045	0·040	0·040*			0·050		
Temp. coeff. (wt-%/°C × 10⁵)	1·2	0·8	—	—			8·9		

* Data for 1550°C

Average result, S_{760} at 1600°C = 0·044 wt-% N

Fe-Al (liquid)

Data for 1600°C:

			References
Wt-% Al	0·25	0·5	115
S_{760} (wt-% N)	0·044	0·044	

Fe-C (liquid)

Data for 1600°C:

Wt-% C	1	2	3	4	5	References
S_{760} (wt-% N)	0·032	0·021	0·015	0·012	—	100
	0·033	0·022	0·015	0·012	—	103
	0·028	0·021	0·013	—	—	114
	0·032	0·024	0·017	0·012	—	118
	0·036	0·028	0·020	—	—	106
	0·028	0·021	0·013	0·008	0·004	107
	0·024	0·015	0·011	—	—	115

Fe-Cr (solid γ)

Temp. (°C)	1000	1100	1200	1300	1400	References
S_{760} (wt-% N) at 4·76% Cr...	0·102	0·079	0·063	0·051	0·034*	120
„ „ 8·67% Cr...	0·286	0·193	0·138	0·102	—	
„ „ 14·10% Cr...	0·96	0·48	0·26	—	—	

* δ phase

Fe–Cr (liquid)

Data for 1600°C:

Wt-% Cr	5	10	15	20	40	60	70	Ref.
S_{700} (wt-% N)	0·07	0·12	0·18	0·29	1·1	—	—	102, 115
	0·07	0·12	0·18	0·30	0·86	2·3	—	103
	—	0·17	—	0·24	1·00	—	3·5	108
	—	0·13	—	0·32	2·30	—	—	111

Data for 1700°C:

Wt-% Cr	10	20	40	50	70	80	90	Ref.
S_{700} (wt-% N)	0·10	0·22	0·75	1·16	2·6	3·5	4·5	108

Fe–Co (liquid)

Data for 1600°C:

Wt-% Co	5	10	20	40	60	80	Ref.
S_{700} (wt-% N)	0·040	0·034	0·028	0·023	0·013	0·010	19, 115, 112, 118
	0·040	0·035	—	—	—	—	

Fe–Cu (liquid)

Data for 1600°C:

Wt-% Cu	2	4	6	8	10	Ref.
S_{700} (wt-% N)	0·044	0·042	0·042	0·040	0·039	112, 118, 115
	0·043	0·041	0·039	0·038	—	

Fe–Mn (solid γ)

Temp. (°C)	1050	1200	1340	Ref.
S_{700} (wt-% N) at 0·43% Mn	0·025	—	—	95
" " 0·53% Mn	0·024	—	—	
" " 1·46% Mn	0·025	—	0·020	
" " 12·98% Mn	0·066	0·046	—	

Fe–Mn (liquid)

Data for 1550°C:

Wt-% Mn	5	10	20	40	60	80	Ref.
S_{700} (wt-% N)	0·044	0·06	0·10	0·23	0·50	0·88	116
	—	0·052	0·087	0·22	0·43	0·80	117

Data for 1600°C:

Wt-% Mn	1	2	5	10	20	Ref.
S_{700} (wt-% N)	—	0·048	0·060	0·074	0·098	108
	0·046	0·048	0·056	—	—	115

Fe–Mo (liquid)

Data for 1600°C:

Wt-% Mo	2	4	6	8	10	Ref.
S_{700} (wt-% N)	0·044	0·047	0·051	0·054	0·056	110, 112, 118
	0·045	0·047	0·051	0·054	0·057	115
	0·046	0·048	0·051	0·054	0·056	

Fe–Nb (liquid)

Data for 1600°C:

Wt-% Nb	2	4	6	8	Ref.
S_{700} (wt-% N)	0·060	0·085	0·11	0·16	115

Fe–Ni (solid)

Temp. (°C)	918	999	1217	Ref.
S_{700} (wt-% N) at 1·01% Ni	0·0283	0·0253	0·0215	119
" " 3·98% Ni	0·0252	0·0219	0·0187	
" " 8·11% Ni	0·0205	0·0186	0·0156	
" " 15·46% Ni	0·0132	0·0125	0·0111	
" " 26·89% Ni	0·0062	0·0062	0·0061	
" " 40·7% Ni	0·0019	0·0023	0·0027	
" " >50% Ni	<0·001	<0·001	<0·001	

TABLE 8. SOLUBILITY OF NITROGEN IN METALS WHICH DISSOLVE LOW CONCENTRATIONS OF THE GAS—continued

Fe-Ni (liquid) — Data for 1600°C:

Wt-% Ni	1	2	5	10	25	50
S_{760} (wt-% N)	0·044	0·043	0·040	0·035	—	—

References: 106, 107, 108, 110, 112, 115, 118, 111

	0·044	0·043	0·039	0·033	0·017	0·0067

References: 115

Fe-O (liquid) — Up to 0·2% O, effect of composition is slight but results of determinations are erratic

Fe-S (liquid) — Up to 0·3% S, effect of composition is slight but results of determinations are erratic

References: 112, 115, 118,

Fe-Si (solid α)

Temp. (°C)	700	800	900	1000	1100
S_{760} (wt-% N) at 2·83% Si	0·0013	0·0016	0·0019	0·0021	0·0024

References: 96

(solid γ)

Temp. (°C)	1000	1050	1100	1200	1300	1350
S_{760} (wt-% N) at 0·20% Si	0·024	0·024	—	0·022	—	0·020
,, ,, 0·58% Si	—	0·022	—	0·020	—	0·018
,, ,, 0·90% Si	0·024	—	0·022	0·021	0·020	—
,, ,, 1·26% Si	0·023	—	0·022	0·021	0·019	—

References: 98, 98, 121, 121

(liquid) — Data for 1600°C:

Wt-% Si	2	4	6	8	10
S_{760} (wt-% N)	0·044	0·032	0·025	0·018	0·014
	0·035	0·026	0·020	0·013	0·010
	0·036	0·029	0·022	0·016	0·012

References: 101, 118, 115

Fe-Sn (liquid) — Data for 1600°C:

Wt-% Sn	2	4	6	8	10
S_{760} (wt-% N)	0·044	0·042	0·041	0·041	0·040
	0·044	0·042	0·041		

References: 112, 115

Fe-Ta (liquid) — Data for 1600°C:

Wt-% Ta	2	4	6	8	10
S_{760} (wt-% N)	0·052	0·060	0·072	0·084	0·098

References: 115

Fe-V (liquid)

Temp. (°C)	1600	1700	1800	1900
S_{760} (wt-% N) at 1% V	0·059	0·059	0·059	0·059
,, ,, 2% V	0·06			
,, ,, 3% V	0·074	0·071	0·070	0·067
	0·07			
,, ,, 5% V	0·088	0·084	0·081	0·079
	0·09			
,, ,, 10% V	0·14	0·129	0·120	0·110
	0·37	0·315	0·280	0·237

References: 110, 115, 110, 115, 110, 115, 110, 115, 110, 115

(solid) — A few experiments in which N_2 was absorbed from N_2-H_2 mixtures by an Fe-0·05% V alloy was reported by TURDOGAN et al.

References: 122

Fe-W (liquid) — Data for 1600°C:

Wt-% W	2	4	6	8	10	12	14
S_{760} (wt-% N)	0·044	0·044	0·045	0·045	0·046	0·046	0·046

References: 115

Fe-Mo-V (liquid) — Data for 1700°C:*

Wt-% Mo	6·5	13·4	29·5	6·0	33·3	24·7	50·0
Wt-% V				6·0	33·1	50·6	25·0
S₇₆₀ (wt-% N)	0·083	0·058	0·041	0·316	0·476	0·762	1·35

* Selected compositions—for more comprehensive data see reference 111

* For data at other temperatures in the ranges 1500–1900°C see reference 110

(References 111, 110)

Co (solid)

1. By Sieverts' method no solubility was detected at temperatures up to 1200°C
2. By Lattice parameter measurements a solubility of 0·6 wt-% N at 600°C was deduced but not supported by direct measurements

(References 123, 124)

Co (liquid) — No solubility detected by Sieverts' method at 1600°C (Reference 19)

Co-Fe — See Fe-Co

Cr (solid)

S_{760} has no meaning at moderate temperatures since Cr_2N forms at sub-atmospheric pressure. Data given below are the terminal solubility, S_T at the nitride dissociation pressure, $P_{nitride}$

Temp. (°C)	1100	1200	1300	1400
$P_{nitride}$ (mm)	1·5 ± 0·5	6·6 ± 0·5	19 ± 1	43·5 ± 0·5
S_T (wt-% N)	0·04	0·09	0·14	0·26

(Reference 125)

Cr (liquid)

Temp. (°C)	1600	1650	1700	1725	1750	1800	1850	1898 +
S_{760} (wt-% N)	4·08	3·90	5·7 / 3·84	3·76	5·3 / 3·54	4·8	4·5	3·2

N.B.—Cr-N alloys have lower m. p. than pure Cr

(References 111, 126)

Cr-Ni — See Ni-Cr

Cr-Si (liquid)

Temp. (°C)	1600	1650	1700	1750
S_{760} (wt-% N) at 1·5% Si	3·83	3·68	3·54	3·08
" " 7·5% Si	1·98	1·89	1·72	1·68
" " 10·0% Si	0·84	0·74	0·69	0·62
" " 20·0% Si	0·33	0·30	0·28	0·26

(Reference 126)

Mo (solid)

Temp. (°C)	950	1000	1050	1100	1150
S_{760} (wt-% N)	0·85	0·56	0·41	0·33	0·26

(Reference 127)

Mn (liquid)

Temp. (°C)	1245 +	1300	1400	1500	1600	1700
S_{760} (wt-% N)	3·4 / 3·06	2·8 / 2·59	1·9 / 1·98	1·6 / 1·56	1·1	1·0

(References 128, 116)

Mn-Fe — See Fe-Mn

Ni (liquid)

1. No solubility detected by Sieverts' method at 1600°C
2. S_{760} (wt-% N) at 1600°C is 0·0025
3. S_{760} (wt-% N) at 1600°C is finite but very small and of the order 0·001

(References 19, 112, 118, 111)

Ni-Cr (liquid) — Data for 1600°C:

Wt-% Cr	10	20	30	40	50	60	70
S_{760} (wt-% N)	0·016	0·068	0·17	0·44	1·0	1·7	2·6

(Reference 111)

Ni-Fe — See Fe-Ni

Si — S_{760} (wt-% N) at m. p. is approximately 0·01 (Reference 129)

S_{760} = Solubility at atmospheric pressure expressed as wt-%. The data may be expressed in units of cc N_2 NTP/100 g by multiplying by the factor 800. Data are quoted from several sources where possible because results are liable to considerable experimental error. + = Indicates melting points.

Alternatively the solubility in equilibrium with a nitride phase may be determined. This has been done for iron on several occasions using calorimetry or internal friction measurements, or by determining the nitrogen contents at which pressure of nitrogen in the gas phase becomes invariant with composition. Data are given in Table 9.

Concentrated solutions

In the solid state certain transition metals dissolve large atomic fractions of nitrogen without abruptly changing their metallic characteristics. At high nitrogen concentrations a second phase may appear which persists to compositions corresponding to simple stoichiometric ratios. These systems are best represented as series of isothermal pressure-concentration curves as for the corresponding metal-hydrogen systems but unfortunately the available data is limited. Some results which have been obtained for Nb and Ta are given in *Figures 13–15*.

The Niobium–Nitrogen System

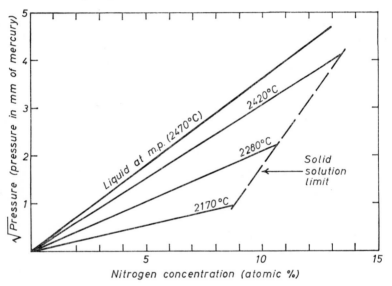

Figure 13. Isothermal pressure-concentration curves for Nb–N (PEMSLER [137])

TABLE 9. SOLUBILITY OF NITROGEN IN IRON IN EQUILIBRIUM WITH NITRIDE PHASES

Solvent	Phase in Equilibrium	Method of Determination	Solubility	References
Pure αFe	Fe₄N	Calorimetry	Temp. (°C): 200, 240, 300, 330, 400, 450, 575 S (wt-% N): 0·008, 0·014, 0·018, 0·026, 0·043, 0·059, 0·097	130
		Invariant pressure	Temp. (°C): 450, 500, 550, 590 S (wt-% N): 0·033, 0·06, 0·070, 0·10	131 93
		Internal friction	Temp. (°C): 250, 300, 350, 400, 450, 500, 575 S (wt-% N): 0·005, 0·008₄, 0·015, 0·025, 0·035, 0·050, 0·075 Temp. (°C): 400, 500, 585* S (wt-% N): 0·025, 0·055, 0·095	132 133
		See also criticism in reference 134 and reply in reference 135		
		* Eutectoid temperature		
	Fe₈N	Internal friction	Temp. (°C): 150, 200, 250, 300 S (wt-% N): 0·0035, 0·010, 0·020, 0·040 Temp. (°C): 20, 100, 200, 300, 400 S (wt-% N): $1·4 \times 10^{-5}$, $5·2 \times 10^{-4}$, 0·0088, 0·055, 0·20	132 133
	N₂ at 1 atm (for comparison)	Internal friction	Temp. (°C): 500, 585, 700, 800, 900 S (wt-% N): $9·0 \times 10^{-4}$, 0·0014, 0·0024, 0·0033, 0·0045 See also Table 8	133
Fe-2·83% Si	Unidentified nitride	Internal friction	Temp. (°C): 300, 400, 500, 600, 700, 800, 900, 1000 S (wt-% N): 0·0010, 0·0015, 0·0024, 0·0040, 0·0061, 0·010, 0·014, 0·019	136

K (II)

The Tantalum–Nitrogen System

Two sets of isothermal pressure-concentration curves for Ta–N have been reported but they are not in exact agreement. They are given in *Figures 14* and *15*.

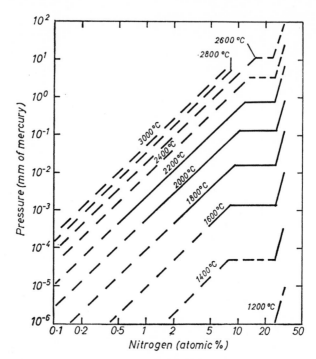

Figure 14.   Tantalum–Nitrogen (GEBHARDT *et al.*[138])
See also: GEBHARDT *et al.*[139]

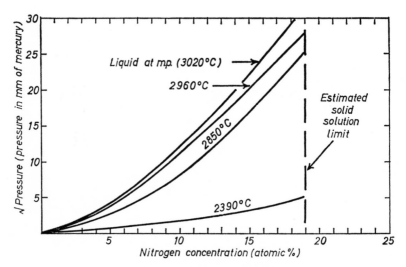

Figure 15.   Tantalum–Nitrogen (PEMSLER [137])

SOLUBILITY OF OXYGEN IN METALS

The free energy of formation of the most stable oxides of most metals is comparatively high and therefore an oxide film is formed on them when exposed to the gas except at very low pressures. The solubility usually required is therefore the terminal solubility in equilibrium with the oxide phase. The oxides of a few metals (Os, Pt, Ru, Au, Ag, Hg, Pd, Rh and Ir) are unstable at atmospheric pressure of oxygen. Of these, the only ones which dissolve appreciable quantities of oxygen are Ag and Pd for which the solubility is conveniently described as concentrations in equilibrium with oxygen at atmospheric pressure. Oxygen is completely insoluble in Hg. Certain transition metals dissolve large quantities of oxygen before a separate oxide phase appears, but the data is not very complete.

Data for the known solubility limits in metal/oxide systems are given in Table 10. In a number of cases these solubilities are small and not easily measured. The usual method of establishing equilibrium by diffusion inwards from a surface oxide phase (or from the gas phase where applicable) is liable to lead to erroneous results unless the metal is very pure and, in particular, free from solute metals which form oxides more stable than that of the solvent metal.

Literature references for certain transition metal–oxygen systems are given following Table 10.

The Niobium–Oxygen System

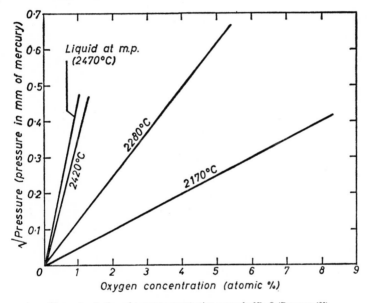

Figure 16. *Isothermal pressure-concentration curves for* Nb–O (PEMSLER [137])

See also: ELLIOT [166]

TABLE 10. SOLUBILITY OF OXYGEN IN METALS

Metal	System	Solubility	References
Ag (solid)	Ag(s)–O₂(g)	Temp. (°C): 200, 300, 400, 500, 600, 700, 800, 900 S_{760} (wt-% O): $1{\cdot}87\times10^{-3}$, $1{\cdot}3\times10^{-3}$, $1{\cdot}16\times10^{-3}$, $1{\cdot}26\times10^{-3}$, $1{\cdot}74\times10^{-3}$, $2{\cdot}55\times10^{-3}$, $4{\cdot}71\times10^{-3}$ $3{\cdot}3\times10^{-6}$, $3{\cdot}0\times10^{-5}$, $1{\cdot}4\times10^{-4}$, $4{\cdot}4\times10^{-4}$, $1{\cdot}07\times10^{-3}$, $2{\cdot}16\times10^{-3}$, $3{\cdot}81\times10^{-3}$, $6{\cdot}14\times10^{-3}$	140 141
(liquid)	Ag(l)–O₂(g)	Temp. (°C): 973, 1024, 1075, 1125 S_{760} (wt-% O): $0{\cdot}305$, $0{\cdot}295$, $0{\cdot}277$, $0{\cdot}264$	140
Co (solid)	αCo(s)–CoO(s)	Temp. (°C): 600, 700, 810, 875 S_T (wt-% O): $0{\cdot}6\times10^{-2}$, $0{\cdot}9\times10^{-2}$, $1{\cdot}6\times10^{-2}$, $2{\cdot}05\times10^{-2}$	142
	βCo(s)–CoO(s)	Temp. (°C): 875, 945, 1000, 1200 S_T (wt-% O): $0{\cdot}58\times10^{-2}$, $0{\cdot}7\times10^{-2}$, $0{\cdot}8\times10^{-2}$, $1{\cdot}3\times10^{-2}$	142
(liquid)	Co(l)–CoO(s)	Temp. (°C): 1550, 1600, 1650 S_T (wt-% O): $0{\cdot}13$, $0{\cdot}16$, $0{\cdot}23$ See also reference 143 for activities of O in Co–Ni alloys	143
Cr (solid)	Cr(s)–Cr₂O₃(s)	Temp. (°C): 1350 S_T (wt-% O): approximately 0·03%	144
Cu (solid)	Cu(s)–Cu₂O(s)	Temp. (°C): 550, 600, 700, 800, 900, 950, 1000, 1050 S_T (wt-% O): $1{\cdot}5\times10^{-3}$, $7{\cdot}1\times10^{-3}$, —, $9{\cdot}4\times10^{-3}$, —, $1{\cdot}00\times10^{-2}$, —, $1{\cdot}56\times10^{-2}$ —, $1{\cdot}6\times10^{-3}$, $1{\cdot}7\times10^{-3}$, $2{\cdot}1\times10^{-3}$, $2{\cdot}7\times10^{-3}$, $3{\cdot}4\times10^{-3}$, $4{\cdot}6\times10^{-3}$, $7{\cdot}7\times10^{-3}$	145 146
Fe (solid)	Fe(s)–FeO(s)	Maximum solubility at 1330–1423°C $0{\cdot}003 \pm 0{\cdot}003$ wt-% O $< 0{\cdot}003$ wt-% O $\sim 0{\cdot}0006$ wt-% O	147 148 149
(liquid)	Fe(l)–FeO(l)	Temp. (°C): 1550, 1600, 1650, 1700 S_T (wt-% O): $0{\cdot}18$, $0{\cdot}23$, $0{\cdot}28$, $0{\cdot}34$ See also references 143, 158 and 159 for activities of O in Fe–Ni and Fe–Co alloys	150, 151, 143
Ge (solid)	Ge(s)–GeO₂(s)	Temp. (°C): 650, 700, 750, 800, 937 S_T (wt-% O): 4×10^{-5}, 8×10^{-5}, $1{\cdot}5\times10^{-4}$, $3{\cdot}0\times10^{-4}$, $1{\cdot}1\times10^{-3}$	152 153
K (liquid)	K(l)–K₂O(s)	Temp. (°C): 100, 150, 200, 250, 300 S_T (wt-% O): $0{\cdot}10$, $0{\cdot}17$, $0{\cdot}27$, $0{\cdot}41$, $0{\cdot}60$	154
Li (liquid)	Li(l)–Li₂O(s)	Temp. (°C): 250, 400 S_T (wt-% O): $0{\cdot}0109$, $0{\cdot}066$	155

Metal	System									Ref.
Na (liquid)	Na(l)–Na₂O(s)	Temp. (°C)	100	200	300	400	500	550		154
		S_{760} (wt-% O)	0·002	0·005	0·010	0·018	0·047	0·08		
Ni (solid)	Ni(s)–NiO(s)	Temp. (°C)	600	800	1000	1200				156
		S_{760} (wt-% O)	0·020	0·019	0·014	0·012				
Ni (liquid)	Ni(l)–NiO(l)	Temp. (°C)	1138*	1450	1500	1550	1600	1650	1700	157
		S_{760} (wt-% O)	0·236	—	—	—	—	—	—	158
			—	0·28	0·40	0·58	0·82	1·15	1·59	159
			—	0·04	0·08	0·15	0·30	0·52	0·93	160
			—	0·28	0·46	0·72	1·10	1·66	—	

See also references 143, 158 and 159 for activities of O in Ni-Fe and Ni-Co alloys

* Eutectic

Metal	System									Ref.
Pb (liquid)	Pb(l)–PbO(s)	Temp. (°C)	350	450	550					161
		S_{760} (wt-% O)	5·4 × 10⁻⁴	8·6 × 10⁻⁴	13·2 × 10⁻⁴					
Pd (solid)	Pd(s)–O₂(g)	Temp. (°C)	1200							162
		S_{760} (wt-% O)	>0·05							
Rh (solid)	Rh(s)–O₂(g)	Slight solubility								162
Si (solid)	Si(s)–SiO₂(s)	Temp. (°C)	1000	1100	1200	1300	1412			163
		S_{760} (wt-% O)	2·8 × 10⁻⁴	5·3 × 10⁻⁴	9·1 × 10⁻⁴	1·5 × 10⁻³	2·2 × 10⁻³			164
Sn (liquid)	Sn(l)–SnO₂(s)	Temp. (°C)	536	600	700	751				165
		S_{760} (wt-% O)	0·00018	0·00055	0·0028	0·0049				

S_{760} = Solubility at atmospheric pressure expressed as wt-%.
S_T = Terminal solubility in equilibrium with an oxide phase.
The data may be expressed in units of cc O₂ NTP/100g by multiplying by the factor 700.
+ indicates melting points.   (s) = solid.   (l) = liquid.   (g) = gas.

The Tantalum–Oxygen System

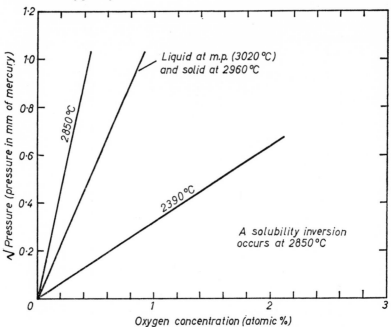

Figure 17. Isothermal pressure-concentration curves for Ta–O *at high temperatures* (PEMSLER [137])
See also: GEBHARDT and SEGHEZZI [168–172, 174]; GEBHARDT and PREISENDANZ [173]; POWERS and DOYLE [175];
(confirmatory data)

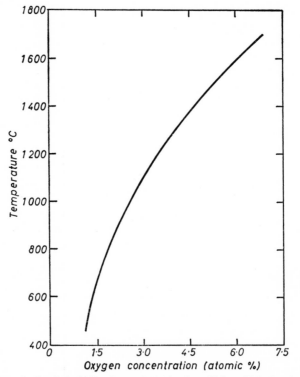

Figure 18.   Terminal solubility of oxygen in tantalum in equilibrium with Ta$_2$O$_5$ *at low temperatures*
(GEBHARDT and SEGHEZZI [167])

The Titanium–Oxygen System

For uptake of oxygen by titanium see: KOFSTAD *et al.*[176]; EHRLICH[177]; McQUILLAN[178]; HURLEN[197]; HEPWORTH and SAMPLE[180]; HEPWORTH and SCHUHMANN[181]

The Vanadium–Oxygen System

See: ALLEN *et al.*[182]; ROSTOKER and YAMAMOTO[183]; GUREVICH and ORMONT[184, 185]

The Zirconium–Oxygen System

See: DE BOER and FAST[186]; KUBASCHEWSKI and DENCH[187]; HOLMBERG and DAGERHAMN[188]; GEBHARDT *et al.*[189]

SOLUTIONS OF INERT GASES IN METALS

The following methods for producing solutions of inert gases in metals have been reported.

1. The solute is introduced by producing an electric discharge at 200–500 V on a hollow cathode of the metal in the gas at low pressure (*e.g.* 0·5 to 2 mm mercury pressure).

2. The solute is injected into thin foil as ions accelerated by a high voltage (*e.g.* 40 kV).

3. The solute is injected as α-particles after cyclotron acceleration to, *e.g.* 38 MeV.

4. Trace quantities of solute are introduced by exposing liquid metal to a neutral gas phase (detected by a radioactive tracer technique).

5. Trace quantities of solute are introduced interstitially into the pseudo metals (*e.g.* Ge or Si) by diffusion from a neutral gas phase.

6. The solute is formed *in situ* by radioactive decay of solvent atoms after activation by neutron irradiation.

7. The solute is formed *in situ* by radioactive decay of atoms of a small alloy addition after activation by neutron irradiation.

8. The solute is formed *in situ* by fission of solvent atoms.

Except in the pseudo metals [207, 44] and possibly also liquid metals,[195] dissolved inert gases cannot be in equilibrium with the gas phase and the quantity of dissolved gas is determined by the technique used to form the solution, *e.g.* by the intensity and duration of the ion current applied to a metal. Systems which have been studied are listed in Table 11.

TABLE 11. REPORTED SOLUTIONS OF INERT GASES IN METALS

System	Method of introducing solute	References	System	Method of introducing solute	References
Ag–A	Method 1	190, 191, 193	Pb–A	Method 2	192
	Method 2	192	Pb–Kr	Method 4	195
Ag–Kr	Method 1	190, 191, 194			
	Method 4	195	Si–He	Method 5	44, 207
Ag–Xe	Method 1	190, 191, 196			
			Sn–Kr	Method 4	195
Al–A	Method 2	192			
Al–He	Method 3	197, 198, 199	Ti–He		208
Al–Li–He	Method 7	200, 201			
Al–U–Kr	Method 7	202	U–A	Method 1	190, 191
			U–Kr	Method 1	190, 191
Au–A	Method 2	192		Method 2	192
				Method 8	209, 210, 211, 212, 213
Be–He	Method 6	203, 204	U–Xe	Method 1	190, 191
	Method 3	205		Method 8	212, 213
Cu–He	Method 3	197, 206			
			Zr–A	Method 1	190, 191
Ge–He	Method 5	44, 207	Zr–Kr	Method 1	190, 191
			Zr–Xe	Method 1	190, 191

REFERENCES

1. R. M. Barrer, " Diffusion in and Through Solids," Cambridge: **1941**.
2. O. Kubaschewski, *Z. Electrochem.*, 1938, **44**/2, 152.
3. J. W. McBain, " Sorption of Gases by Solids," London: **1932**.
4. A. Nikuradse and R. Ulbricht, " Das Zweistoffsystem Gas–Metal," Munich: **1950**.
5. A. Sieverts, *Z. Metallkunde*, 1929, **21**, 37.
6. C. J. Smithells, " Gases and Metals," London: **1937**.

7. C. R. Cupp, *Progress in Metal Physics*, **4**, Pergamon Press, (1954), 105.
8. D. P. Smith, " Hydrogen in Metals," Chicago: **1948**.
9. H. Kostron, *Z. Metallkunde*, 1952, **43**, 269 and 373.
10. P. Cotterill, *Progress in Materials Science*, **9**, Pergamon Press (1961), 205.
11. R. Fowler and C. J. Smithells, *Proc. Roy. Soc.*, 1937, **160**, 37.
12. R. M. Barrer, *Discussions Faraday Soc.*, **4**, 1948, 68.
13. D. E. Rimmer and A. H. Cottrell, *Phil. Mag.*, 1957, **23**, 1345.
14. E. W. R. Steacie and F. M. G. Johnson, *Proc. Roy. Soc.*, 1928A, **117**, 662.
15. C. E. Ransley and H. Neufeld, *J. Inst. Metals*, 1948, **74**, 599.
16. W. Eichenauer, K. Hattenbach and A. Pebler, *Z. Metallkunde*, 1961, **52**, 682.
17. W. R. Opie and N. J. Grant, *Trans. Amer. Inst. Min. Met. Eng.*, 1950, **188**, 1237.
18. A. Sieverts and H. Hagen, *Z. physical. Chem.*, 1934, **169**, 237.
19. T. Busch and R. A. Dodd, *Trans. Met. Soc. Amer. Inst. Min. Met. Eng.*, 1960, **218**, 488.
20. E. Martin, *Arch. Eisenhüttenwesen*, 1929/30, **3**, 407.
21. L. Luckmeyer-Hasse and H. Schenk, *Arch. Eisenhüttenwesen*, 1932–3, **6**, 209.
22. W. Eichenauer and A. Pebler, *Z. Metallkunde*, 1957, **48**, 373.
23. A. Sieverts, *Z. Physical. Chem.*, 1911, **77**, 611.
24. P. Röntgen and F. Möller, *Metallwirtschaft*, 1934, **13**, 81, 97.
25. M. B. Bever and C. F. Floe, *Trans. Amer. Inst. Min. Met. Eng.*, 1944, **156**, 149.
26. R. Eborall and A. J. Swain, *J. Inst. Metals*, 1952–3, **81**, 497.
27. W. Eichenauer, H. Kunzig and A. Pebler, *Z. Metallkunde*, 1958, **49**, 220.
28. A. Sieverts, G. Zapf and H. Moritz, *Z. physical Chem.*, 1938, **183**, 19.
29. M. H. Armbruster, *J. Amer. Chem. Soc.*, 1943, **65**, 1043.
30. F. de Kazinczy and O. Lindberg, *Jernkontorets Ann.*, 1960, **144**, 288.
31. H. Schenck and H. Wünsch, *Arch. Eisenhüttenwesen*, 1961, **32**, 779.
32. H. Liang, M. B. Bever and C. F. Floe, *Trans. Amer. Inst. Min. Met. Eng.*, 1946, **167**, 395.
33. M. M. Karnaukhov and A. N. Morazov, *Isvest. Akad. Nauk (Otdelenie Tekh. Nauk)*, December 1948, 1845.
34. H. Winterhager, *Aluminium-Archiv.*, 1938, **12**, 7.
35. R. S. Busk and E. G. Bobalek, *Trans. Amer. Inst. Min. Met. Eng.*, 1947, **171**, 261.
36. F. Sauerwald, *Zeit. anorg. Chem.*, 1949, **258**, 27.
37. J. Koenman and A. G. Metcalf, *Trans. Amer. Soc. Metals*, 1959, **51**, 1072.
38. A. Sieverts and H. Moritz, *Z. physical. Chem.*, 1938, 180/4, 249.
39. E. V. Potter and H. C. Lukens, *Trans. Amer. Inst. Min. Met. Eng.*, 1947, **171**, 401.
40. W. R. Opie and N. J. Grant, *Trans. Amer. Inst. Min. Met. Eng.*, 1951, **191**, 244.
41. W. Mannchen and R. Bauman, *Metall.*, 1955, **9**, 686.
42. W. Hofmann and J. Maatsch, *Z. Metallkunde*, 1956, **47**, 89.
43. A. Sieverts, *Ber. deut. chem. Ges.*, 1912, **45**, 221.
44. A. van Wieringen and N. Warmoltz, *Physica*, 1956, **22**, 849.
45. D. T. Peterson and M. Indig, *J. Amer. Chem. Soc.*, 1960, **82**, 5645.
46. W. D. Treadwell and J. Stecher, *Helv. Chim. Acta*, 1953, **36**, 1820.
47. W. C. Johnson, M. F. Stubbs, A. E. Sidwell and A. Pechukas, *J. Amer. Chem. Soc.*, 1939, **61**, 318.
48. D. T. Peterson and V. G. Fattore, *J. Phys. Chem.*, 1961, **65**(11), 2062.
49. J. F. Stampfer, C. E. Holley and J. F. Shuttle, *J. Amer. Chem. Soc.*, 1960, **82**, 3504.
50. J. A. Kenneley, J. W. Varwig and H. W. Myers, *J. Phys. Chem.*, 1960, **64**(5), 703.
51. M. W. Mallet and M. J. Trzeciak, *Trans. Amer. Soc. Metals*, 1958, **50**, 981.
52. H. C. Mattrow, *J. Phys. Chem.*, 1955, **59**, 93.
53. R. N. R. Mulford and C. E. Holley, *J. Phys. Chem.*, 1955, **59**, 1222.
54. A. Sieverts and G. Muller-Goldegg, *Z. anorg. allgem. Chem.*, 1923, **131**, 65.
55. —— and E. Roell, *Z. anorg. Chem.*, 1925, **146**, 149.
56. —— and A. Gotta, *ibid.*, 1928, **172**, 1.
57. L. Espagno, P. Azou and P. Bastien, *Compt. Rend.*, 1960, **250**, 4352.
58. —— *Mem. Sci. Rev. Met.*, 1962, **59**, 182.
59. W. M. Albrecht, W. D. Goode and M. W. Mallet, *J. Electrochem. Soc.*, 1959, **106**, 981.
60. —— *ibid.*, 1958, **105**, 219.
61. S. Komjathy, *J. Less Common Metals*, 1960, **2**, 466.
62. A. Sieverts and E. Roell, *Z. anorg. allgem. Chem.*, 1926, **150**, 261.
63. P. L. Levine and K. E. Weal, *Trans. Faraday Soc.*, 1960, **56**, 357.
64. L. J. Gillespie and F. P. Hall, *J. Amer. Chem. Soc.*, 1926, **48**, 1207.
65. —— and L. S. Galstaun, *J. Amer. Chem. Soc.*, 1936, **58**, 2565.
66. T. S. Perminov, A. A. Orlov and A. N. Frumkin, *Doklady Akad. Nauk, SSSR*, 1952, **84**, 749.
67. D. H. Everett and P. Nordon, *Proc. Roy. Soc.*, 1960A, **259**, 341.
68. T. B. Flanagan, *J. Phys. Chem.*, 1961, **65**(2), 280.
69. I. E. Nakhutin and E. I. Sutyagina, *Fizica Metallov i Metallovedenie*, 1959, **7**, 459.
70. P. Mitacek and J. G. Aston, *J. Amer. Chem. Soc.*, 1963, **85**(2), 137.
71. A. W. Carson, T. B. Flanagan and F. A. Lewis, *Trans. Faraday Soc.*, 1960, **56**, 1332 and 371.
72. R. A. Karpova and I. P. Tverdovsky, *Zhur. Fiz. Khim*, 1959, **33**, 1393.
73. R. N. R. Mulford and G. Sturdy, *J. Amer. Chem. Soc.*, 1955, **77**, 3449.
74. M. W. Mallet and B. G. Khoehl, *J. Electrochem. Soc.*, 1962, **109**, 611 and 968.
75. P. Kofstad, W. E. Wallace and L. J. Hyvönen, *J. Amer. Chem. Soc.*, 1959, **81**, 5015.
76. M. W. Mallet and I. E. Campbell, *J. Amer. Chem. Soc.*, 1951, **73**, 4850.
77. D. T. Peterson and D. G. Westlake, *Trans. Met. Soc. Amer. Inst. Min. Met. Eng.*, 1959, **215**, 445.
78. A. D. McQuillan, *Proc. Roy. Soc.*, 1950A, **204**, 309.
79. G. A. Lenning, C. M. Craighead and R. I. Jaffee, *Trans. Amer. Inst. Min. Met. Eng.*, 1954, **200**, 367.
80. A. D. McQuillan, *J. Inst. Metals*, 1951, **79**, 73.
81. P. Kofstad and W. E. Wallace, *J. Amer. Chem. Soc.*, 1959, **81**, 5019.
82. C. E. Ells and A. D. McQuillan, *J. Inst. Metals*, 1956–7, **85**, 89.
83. J. Motz, *Z. Metallkunde*, 1962, **53**, 770.
84. C. M. Schwartz and M. W. Mallet, *Trans. Amer. Soc. Metals*, 1954, **46**, 640.
85. E. A. Gulbransen and K. F. Andrew, *J. Electrochem. Soc.*, 1954, **101**, 474, and *J. Metals*, 1955, **7**, 136.
86. R. F. Edwards, P. Levesque and D. Cubicciotti, *J. Amer. Chem. Soc.*, 1955, **77**, 1307.
87. M. W. Mallet and W. M. Albrecht, *J. Electrochem. Soc.*, 1957, **104**, 142.
88. L. Espagno, P. Azou and P. Bastien, *Mem. Sci. Rev. Met.*, 1960, **57**, 254.
89. G. G. Libowitz, *J. Nuclear Materials*, 1962, **5**, 228.
90. L. D. La Grange, L. J. Dijkstra, J. M. Dixon and U. Merten, *J. Phys. Chem.*, 1959, **63**, 2035.
91. C. C. Addison, R. J. Pulham and R. J. Roy, *J. Chem. Soc.*, 1965, 116.
92. D. D. Williams, J. A. Grand and R. R. Miller, *J. Phys. Chem.*, 1957, **61**, 379.
93. H. A. Wriedt and L. S. Darken, *Trans. Met. Soc. Amer. Inst. Min. Met. Eng.*, 1965, **233**, 111.
94. —— *ibid.*, 1965, **233**, 122.
95. A. Sieverts and G. Zapf, *Z. phys. Chem.*, 1935, **174**, 359.
96. N. S. Corney and E. T. Turkdogan, *J. Iron Steel Inst.*, 1955, **180**, 344.
97. A. Sieverts, G. Zapf and H. Moritz, *Z. phys. Chem.*, 1938, **183**, 19.
98. L. S. Darken, R. P. Smith and C. W. Filer, *Trans. Amer. Inst. Min. Met. Eng.*, 1951, **191**, 1174.
99. J. Chipman and D. Murphy, *Trans. Amer. Inst. Min. Met. Eng.*, 1935, **116**, 179.
100. L. Eklund, *Jernkontorets Ann.*, 1939, **123**, 545.
101. J. C. Vaughan and J. Chipman, *Trans. Amer. Inst. Min. Met. Eng.*, 1940, **140**, 224.
102. R. M. Brick and J. A. Creevy, *Metals Tech. A.I.M.E. Tech. Pub.*, No. 1165, April 10, 1940.

103. T. Kootz, *Arch. Eisenhüttenwesen*, 1941, **15**, 77.
104. C. R. Taylor and J. Chipman, *Trans. Amer. Inst. Min. Met. Eng.*, 1943, **154**, 228.
105. M. M. Karnaukov and A. M. Morozov, *Bull. acad. Sci. U.R.S.S. Classe sci. tech.*, 1947, **735**, Brutcher Transl. no. 2029.
106. T. Saito, *Sci. Repts. Research Insts. Tohoku Univ.*, Ser. A., 1949, **1**, 411.
107. ——, *ibid.*, 419.
108. H. Wentrup and O. Reif, *Arch Eisenhüttenwesen*, 1949, **20**, 359.
109. Y. Kasamatu and S. Matoba, *Technol. Repts. Tohoku Univ.*, 1957, **22**, No. 1.
110. V. Kashyap and N. Parlee, *Trans. Amer. Inst. Min. Met. Eng.*, 1958, **212**, 86.
111. J. Humbert and J. F. Elliot, *Trans. Met. Soc. Amer. Inst. Min. Met. Eng.*, 1960, **218**, 1076.
112. H. Schenck, M. Frohberg and H. Graf, *Arch. Eisenhüttenwesen*, 1958, **29**, 673.
113. V. P. Fedotov and A. M. Samarin, *Doklady Acad. Nauk SSSR*, 1958, **122**, 597.
114. S. Maekawa and Y. Nakagawa, *Tetsu to Hagane*, *J. Iron Steel Inst.*, Japan, 1959, **45**, 255.
115. R. D. Pehlke and J. F. Elliott, *Trans. Met. Soc. Amer. Inst. Min. Met. Eng.*, 1960, **218**, 1088.
116. S. Z. Beer, *Trans. Met. Soc. Amer. Inst. Min. Met. Eng.*, 1961, **221**, 2.
117. R. A. Dodd and N. A. Gokcen, *Trans. Met. Soc. Amer. Inst. Min. Met. Eng.*, 1961, **221**, 233.
118. H. Schenck, M. Frohberg and H. Graf, *Arch. Eisenhüttenwesen*, 1959, **30**, 533.
119. H. A. Wriedt and O. D. Gonzalez, *Trans. Met. Soc. Amer. Inst. Min. Met. Eng.*, 1961, **221**, 532.
120. E. T. Turkdogan and S. Ignatowicz, *J. Iron Steel Inst.*, 1958, **188**, 242.
121. —— ——, *J. Iron Steel Inst.*, 1957, **185**, 200.
122. —— and J. Pearson, *J. Iron Steel Inst.*, 1955, **181**, 227.
123. A. Sieverts and H. Hagen, *Z. physical Chem.*, 1934, **169**, 337.
124. R. Juza and W. Sachsze, *Z. anorg. Chem.*, 1945, **253**, 95.
125. A. U. Seybolt and R. A. Oriani, *Trans. Amer. Inst. Min. Met. Eng.*, 1956, **206**, 556.
126. V. S. Mozgovoy and A. M. Samarin, *Doklady. Akad. Nauk SSSR*, 1950, **74**, 729.
127. A. Sieverts and H. Brunig, *Arch. Eisenhüttenwesen*, 1933, **7**, 641.
128. N. A. Gokcen, *Trans. Met. Soc. Amer. Inst. Min. Met. Eng.*, 1961, **221**, 200.
129. W. Kaiser and C. D. Thurmond, *J. Appl. Phys.*, 1959, **30**, 427.
130. G. Borelius, S. Berglund and O. Avsan, *Arkiv. Fysick*, 1950, **2**, 551.
131. V. G. Pararjpe, M. Cohen, M. B. Bever and C. F. Floe, *Trans. Amer. Inst. Min. Met. Eng.*, 1950, **188**, 261.
132. L. J. Dijkstra, *Trans. Amer. Inst. Min. Met. Eng.*, 1949, **185**, 252.
133. J. D. Fast and M. B. Verrijp, *J. Iron Steel Inst.*, 1955, **180**, 337.
134. H. U. Aström and G. Borelius, *Acta Met.*, 1954, **2**, 547.
135. J. D. Fast and M. B. Verrijp, *Acta Met.*, 1955, **3**, 203.
136. D. A. Leak, W. R. Thomas and G. M. Leak, *Acta Met.*, 1956, **3**, 501.
137. J. P. Pemsler, *J. Electrochem. Soc.*, 1961, **108**, 744.
138. E. Gebhardt, H. D. Seghezzi and E. Fromm, *Z. Metallkunde*, 1961, **52**, 464.
139. —— and W. Dürrschnabel, *Z. Metallkunde*, 1958, **49**, 577.
140. E. W. R. Steacie and F. M. G. Johnson, *Proc. Roy. Soc.*, 1926, **A112**, 542.
141. W. Eichenauer and G. Muller, *Z. Metallkunde*, 1962, **53**, 321.
142. A. U. Seybolt and C. H. Mathewson, *Trans. Amer. Inst. Min. Met. Eng.*, 1935, **117**, 156.
143. V. V. Averin, A. Yu. Polyakov and A. M. Samarin, *Izvest. Akad. Nauk, SSSR*, 1957, (Tekhn.) **(8)**, 120.
144. D. Caplan and A. A. Burr, *Trans. Amer. Inst. Min. Met. Eng.*, 1955, **203**, 1052.
145. F. N. Rhines and C. H. Mathewson, *Trans. Amer. Inst. Min. Met. Eng.*, 1934, **111**, 337.
146. A. Phillips and E. N. Skinner, *Trans. Amer. Inst. Min. Met. Eng.*, 1941, **143**, 301.
147. J. A. Kitchener, J. O'M. Bockris, M. Gleiser and J. W. Evans, *Trans. Faraday Soc.*, 1952, **48**, 955.
148. F. Wever, W. A. Fischer and H. Engelbrecht, *Stahl u. Eisen*, 1954, **74**, 1521.
149. R. Sifferlen, *Comptes Rend., Acad. Sci.*, Paris, 1957, **244**, 1192.
150. J. Chipman and K. L. Fetters, *Trans. Amer. Soc. Metals*, 1941, **29**, 953.
151. C. R. Taylor and J. Chipman, *Trans. Amer. Inst. Min. Met. Eng.*, 1943, **154**, 228.
152. W. Kaiser and C. D. Thurmond, *J. Appl. Physics*, 1961, **32**, 115.
153. E. S. Candidus and D. Tuomi, *J. Chem. Physics*, 1955, **23**, 588.
154. D. D. Williams, J. A. Grand and R. R. Miller, *J. Phys. Chem.*, 1959, **63**, 68.
155. E. E. Hoffman, *Amer. Soc. Test. Mat. Symposium on Newer Materials*, 1959, **1960**, 195.
156. A. U. Seybolt, *Dissertation*, Yale University, 1936.
157. P. D. Merica and R. G. Waltenberg, *Trans. Amer. Inst. Min. Met. Eng.*, 1925, **71**, 715.
158. H. A. Wriedt and J. Chipman, *Trans. Amer. Inst. Min. Met. Eng.*, 1955, **203**, 477.
159. A. M. Samarin and V. P. Fedotov, *Izvest. Akad. Nauk, SSSR*, 1956, (Tekhn.) **(6)**, 119.
160. J. E. Bowers, *J. Inst. Metals*, 1961–2, **90**, 321.
161. K. W. Groshelm-Krisko, W. Hoffman and H. Hanemann, *Z. Metallkunde*, 1944, **36**, 91.
162. E. Raub and N. Plate, *Z. Metallkunde*, 1957, **48**, 529.
163. H. J. Hrostowski and R. H. Kaiser, *Phys. Chem. Solids.*, 1959, **9**, 214.
164. W. Kaiser and P. H. Keck, *J. Appl. Phys.*, 1957, **28**, 1427.
165. T. N. Belford and C. B. Alcock, *Trans. Faraday Soc.*, 1965, **61**, 443.
166. R. P. Elliot, *Amer. Soc. Metals*, Reprint, 1959, (143).
167. E. Gebhardt and H. D. Seghezzi, *Z. Metallkunde*, 1959, **50**, 521.
168. —— ——, *ibid.*, 1959, **50**, 248.
169. —— ——, *ibid.*, 1955, **46**, 560.
170. —— ——, *ibid.*, 1957, **48**, 430.
171. —— ——, *ibid.*, 1957, **48**, 503.
172. —— ——, *ibid.*, 1957, **48**, 559.
173. —— and H. Preisendanz, *Plansee Proc.*, 1955, 254.
174. —— and H. D. Seghezzi, *ibid.*, 1959, 280.
175. R. J. Powers and M. V. Doyle, *Trans. Met. Soc. Amer. Inst. Min. Met. Eng.*, 1959, **215**, 655.
176. P. Kofstad, P. B. Anderson and O. J. Krudtaa, *J. Less Common Metals*, 1961, **3**, 89.
177. F. Ehrlich, *Z. anorg. Chem.*, 1941, **24**, 53.
178. M. K. McQuillan, *Corrosion et Anticorrosion*, 1962, **10**, 361.
179. T. Hurlen, *J. Inst. Metals*, 1960–1, **89**, 128.
180. M. T. Hepworth and W. B. Sample, *Trans. Met. Soc. Amer. Inst. Min. Met. Eng.*, 1962, **224**, 875.
181. M. T. Hepworth and R. Schuhmann, *Trans. Met. Soc. Amer. Inst. Min. Met. Eng.*, 1962, **224**, 928.
182. N. P. Allen, O. Kubaschewski and O. V. Goldbeck, *J. Electrochem. Soc.*, 1951, **98**, 417.
183. W. Rostoker and A. S. Yamamoto, *Trans. Amer. Soc. Metals*, 1955, **47**, 1002.
184. M. A. Gurevich and B. F. Ormont, *Zhur. Neorg. Khim.*, 1957, **2**, 1566, 2581.
185. —— ——, *ibid.*, 1958, **3**, 403.
186. H. J. de Boer and J. D. Fast, *Rec. Trav. chim. Pays-Bas*, 1936, **55**, 449.
187. O. Kubaschewski and W. A. Dench, *J. Inst. Metals*, 1955–6, **84**, 440.
188. B. Holmberg and T. Dagerhamn, *Acta Chem. Scand.*, 1961, **15**, 915.
189. E. Gebhardt, H. D. Seghezzi and W. Durrschnabel, *J. Nuclear Materials*, 1961, **4**, 241, 255 and 269.
190. G. Brebec, V. Lévy and Y. Adda, *Compt. Rend.*, 1961, **252**, 722.
191. V. Lévy *et al.*, *Compt. Rend.*, 1961, **252**, 876.
192. C. W. Tucker and F. J. Norton, *J. Nuclear Materials*, 1960, **2**, 329.
193. A. D. Le Claire and A. H. Rowe, *Rev. Mét.*, 1955, **52**, 94.
194. J. M. Tobin, *Acta Met.*, 1957, **5**, 398.
195. G. W. Johnson and R. Shuttleworth, *Phil. Mag.*, 1959, **4**, 957.
196. J. M. Tobin, *Acta Met.*, 1959, **7**, 701.
197. R. S. Barnes, *Phil. Mag.*, 1960, **5**, 635.

198. C. E. Ells and C. E. Evans, Atomic Energy of Canada Ltd., Rep. 1959 (C R Met-863).
199. ——, *J. Nuclear Materials*, 1962, **5**, 147.
200. G. T. Murray, *J. Appl. Phys.*, 1961, **32**, 1045.
201. D. W. Lillie, *Trans. Met. Soc. Amer. Inst. Min. Met. Eng.*, 1960, **218**, 270.
202. M. B. Reynolds, *Nuclear Sci and Eng.*, 1958, **3**, 428.
203. C. E. Ells and E. C. W. Perryman, *J. Nuclear Materials*, 1959, **1**, 73.
204. V. Lévy, *Bull. Inform. Sci. Tech.*, 1962, **62**, 56.
205. R. S. Barnes and G. B. Redding, *J. Nucl. Energy*, A, 1959, **10**, 32.
206. —— —— and A. H. Cottrell, *Phil. Mag.*, 1958, **3**, 97.
207. A. van Wieringen, Symposium: " La Diffusion dans les Metaux," **1957**, 107.
208. A. M. Rodin and V. V. Surenyants, *Fizika Metallov i Metallovedenie*, 1960, **10**, 216.
209. M. B. Reynolds, *Nuclear Sci. and Eng.*, 1956, **1**, 374.
210. J. F. Walker, U.K. Atomic Energy Authority Publ. 1959 (IGR-TN/W-1046).
211. F. J. Norton, *J. Nuclear Materials*, 1960, **2**, 350.
212. D. L. Gray, U.S. Atomic Energy Commission Rep. 1960 (HW-62639).
213. N. R. Chellew and R. K. Steunenberg, *Nuclear Sci. Eng.*, 1962, **14**, 1.

DIFFUSION IN METALS

INTRODUCTION

In an isotropic medium the diffusion coefficient D^i of species i is defined through Fick's first law,

$$J^i = -D^i \operatorname{grad} c^i \qquad . \qquad . \qquad . \qquad . \qquad . \qquad (1)$$

J^i is the instantaneous net flux of species i, or diffusion current per unit area, and grad c^i is the gradient of the concentration c^i of i. If J and c are measured in terms of the same unit of quantity (e.g. J in g/cm^2 sec, c in g/cm^3), D has the dimensions $[L^2 T^{-1}]$. It is usually expressed as cm^2/sec. Generally, D depends on the concentration.

That matter is to be conserved at each point leads to Fick's second law,

$$\frac{\partial c^i}{\partial t} = \operatorname{div}(D^i \operatorname{grad} c^i) \qquad . \qquad . \qquad . \qquad . \qquad (2)$$

giving the rate of the change of concentration with time to which diffusion gives rise.

The fluxes J^i are referred, at least for practical purposes, to axes fixed in the volume of the sample; but volume changes which take place as a result of diffusion lead to some ambiguity in the definition of such axes. Means have been proposed [4, 11] for avoiding this by using axes scaled to the volume changes, but little use is made of these and it is more usual in accurate work to restrict the range of concentration employed so that volume changes are small or negligible.

When the concentration varies along only one direction, say the x axis, (1) and (2) become

$$J^i = -D^i \frac{\partial c^i}{\partial x} \qquad . \qquad . \qquad . \qquad . \qquad (3)$$

$$\frac{\partial c^i}{\partial t} = \frac{\partial}{\partial x}\left(D^i \frac{\partial c^i}{\partial x}\right) \qquad . \qquad . \qquad . \qquad . \qquad (4)$$

If, furthermore, D is independent of composition, and so also of position in the sample, (4) becomes

$$\frac{\partial c^i}{\partial t} = D^i \frac{\partial^2 c^i}{\partial x^2} \qquad . \qquad . \qquad . \qquad . \qquad (5)$$

In anisotropic media diffusion rates vary with direction. In general, the diffusion flux is in the same direction as grad c *only* when grad c is along one of a set of orthogonal axes known as the 'principal axes of diffusion'. (These always coincide with axes of crystallographic symmetry so there is no difficulty in identifying them, except in cases of symmetry lower than orthorhombic.) For diffusion along principal axes equations like (3) may still be written

$$\left.\begin{array}{l} J_x{}^i = -D_x{}^i(\partial c^i/\partial x) \\ J_y{}^i = -D_y{}^i(\partial c^i/\partial y) \\ J_z{}^i = -D_z{}^i(\partial c^i/\partial z) \end{array}\right\}$$

D_x, D_y and D_z are called 'principal coefficients of diffusion'.

In general grad c and J are not in the same direction. However, if l, m, n are the direction cosines of grad c then a diffusion coefficient for this direction may be defined as the ratio of the component of J along (l, m, n), divided by grad c. This is

$$D_{lmn} = l^2 D_x + m^2 D_y + n^2 D_z \qquad . \qquad . \qquad . \qquad (6)$$

Thus anisotropic diffusion can be completely described in terms of the three principal diffusion coefficients. In uniaxial crystals (tetragonal, trigonal, hexagonal) symmetry dictates, if the z axis is the unique axis, that $D_z = D_y$. Thus D is the same for all directions perpendicular to the unique axis and is often denoted $D_\perp$. D_z is then denoted as $D_\parallel$. (6) may then be written

$$D_\theta = \sin^2\theta \,.\, D_\perp + \cos^2\theta \,.\, D_\parallel \quad . \qquad . \qquad . \qquad . \quad (7)$$

where $\cos\theta \equiv n$.

Equations (4) and (5) still hold for anisotropic diffusion, with D given by (6) and (7).

Equation (1) provides a formal definition of a diffusion coefficient as the ratio of J^i to grad c^i. It also assumes that J^i is determined only by grad c^i. In the very large majority of diffusion measurements that have been made this holds true so that the above simple equations provide an adequate description of the diffusion process taking place. Such measurements are of three main types and these are discussed first and the nature of the diffusion coefficients they entail. They are:

1. Measurements which entail diffusion under a chemical concentration gradient (Chemical Diffusion Measurements—Table 4).

(i) Diffusion of a single interstitial solute into a pure metal.
(ii) Interdiffusion of two metals which form substitutional solid solutions (or interdiffusion between two alloys of the two metals).

2. Measurements which entail diffusion in essentially chemically homogeneous systems. These are possible through the use of radioactive tracers.

The diffusion of an interstitial solute in a pure metal [1(i)] is described by a single equation like (1) and the D has a simple and well-defined physical significance as describing diffusion of solute relative to the solvent lattice.

The same is true for the D for diffusion into a metal or alloy of any radioactive tracer. The methods employed (see below) require such extremely small amounts and gradients of tracer that the system remains chemically homogeneous during diffusion. Any diffusion of other constituents is altogether negligible so that D refers simply to the diffusion of the tracer species relative to the solvent lattice.

For the interdiffusion of two metals or alloys [1(ii)] the situation is a little less simple. There would appear to be two diffusion coefficients required, one for each species, but *referred to volume fixed axes* these are equal because grad $c_1 = -\text{grad}$ c_2 and J_1 must be equal and opposite to J_2. Again a single equation like (1) suffices to describe the diffusion process and the single D refers to the diffusion rate of either species relative to these axes. It is called the *chemical interdiffusion coefficient* and usually denoted $\tilde{D}$ (Table 4).

For many practical purposes $\tilde{D}$ is an adequate measure of the diffusion behaviour of a binary substitutional system. But of more fundamental physical interest are the rates of diffusion of the two species relative to local lattice planes. It is well established that generally these rates are not equal in magnitude. There is therefore a net total flux of atoms across any lattice plane, and if the density of lattice sites is to be conserved each plane in the diffusion zone must shift to compensate for this imbalance of the fluxes across it. At the same time lattice sites are created on one side of the sample and eliminated at the other, processes which are achieved by the creation and annihilation of vacancies. This shift of lattice planes, known as the Kirkendall Effect, is observed experimentally as a movement of inert markers, usually fine insoluble wires, incorporated into the sample before diffusion. It is clear then that diffusion occurs on a lattice which locally is moving relative to the axes with respect to which $\tilde{D}$ was calculated. To provide a more complete description of binary substitutional diffusion it is therefore necessary to introduce diffusion coefficients D_A and D_B to describe diffusion of the two species relative to lattice planes. It is easy to show that these are related to D by the equation

$$\tilde{D} = N_A D_B + N_B D_A \quad . \qquad . \qquad . \qquad . \quad (8)$$

where N_A and N_B are the fractional concentrations of A and B. D_A and D_B, which are of more direct physical interest than D, are known as the *intrinsic* or *partial chemical diffusion coefficients*.

The velocity v of a marker is given by

$$v = (D_A - D_B)\partial N_A/\partial x, \quad . \qquad . \qquad . \qquad . \quad (9)$$

where $\partial N_A/\partial x$ is the concentration gradient at the marker; so in principle D_A and D_B can be calculated separately when $\tilde{D}$ and v have been measured. In practice this is done usually only for markers placed at the original interface between the two interdiffusing metals or alloys: in this case a measurement of the *displacement* x_m of the marker after time t allows v to be obtained simply, for $v = x_m/2t$.

Equations (8) and (9) assume no net volume change and a compensation of the flux difference which is complete and which occurs by bulk motion along only the diffusion direction. These conditions are rarely met fully in practice, as is seen from the occurrence often of lateral changes in dimensions and of a porosity in the *side* of the diffusion zone suffering a net loss of atoms. This porosity, attributed to vacancies precipitating instead of being eliminated at sinks, suggests abnormal vacancy concentrations may be present in the diffusion zone. Because it is difficult to take into account the effect these abnormal conditions in the diffusion zone may have on the calculated values of $\tilde{D}$ and v, and hence on D_A and D_B, chemical inter-diffusion experiments may provide results of limited accuracy and, for theoretical purposes, of limited significance: their effect is of course smaller the smaller the concentration gradients employed.

By contrast, radioactive tracer methods altogether avoid these difficulties and uncertainties associated with diffusion in a chemical gradient, and so are preferred in any investigation with a theoretical objective. They have the further advantage that the diffusion coefficients of the several species of an alloy can be determined separately and directly, rather than through any composite coefficient like $\tilde{D}$. These are referred to as *tracer diffusion coefficients* (Table 3) and will be denoted D_A^*, D_B^* etc. to distingusih them from the partial chemical diffusion co-efficients D_A and D_B determined by chemical diffusion methods.†

Except at vanishingly small concentrations of A, D_A and D_A^* differ fundamentally because the presence of the chemical concentration gradient under which D_A is measured imposes on the otherwise random motion of the atoms a bias, which makes atoms jump preferentially in one direction along the concentration gradient. Simple thermodynamic considerations lead to the relation

$$D_A = D_A^* \left(1 + \frac{\partial \ln \gamma_A}{\partial \ln N_A} \right) \qquad \cdot \qquad \cdot \qquad \cdot \qquad \cdot \quad (10)$$

between a partial chemical D_A and the corresponding tracer D_A^* measured at the same concentration. γ_A is the activity coefficient of A. In a binary system the bracket term is the same for both species (Gibbs–Duhem relation). Thus

$$\frac{D_A}{D_A^*} = \frac{D_B}{D_B^*} \qquad \cdot \qquad \cdot \qquad \cdot \qquad \cdot \quad (11)$$

(10) and (11) are approximate forms of more elaborate theoretical expressions, but are reasonably well obeyed experimentally.

When D_A^* is measured at the extremely small concentrations of A that tracer methods permit (by diffusion of tracer into pure metal B) it is called the *tracer impurity diffusion coefficient* of A in B (Table 2). Such coefficients are of especial theoretical interest because of the particularly simple type of diffusion they describe (see footnote).

Finally, tracer methods are used as the commonest means of measuring *self diffusion coefficients* in pure metals (Tables 1, 5 and 6). By self-diffusion is meant of course the diffusion of a species in the pure lattice of its own kind.

For chemical diffusion in systems of more than two components, equation (1) and those following are inadequate. Experimentally it is found that when three or more components are present a concentration gradient of one species can lead to a diffusion flow of another, even if this is distributed homogeneously to start with. To cater for such cases Fick's first law is generalized by writing

$$J_i = \sum_{j=1}^{N} D_{ij} \frac{\partial c_j}{\partial x} \ (j = 1, 2 \dots N) \qquad \cdot \qquad \cdot \qquad \cdot \quad (12)$$

But if there are n interstitial and $N-n$ substitutional components, and if the J_i are referred to volume-fixed axes then the relations $\sum_{j=n+1}^{N} J_i = 0$ and $\sum_{j=n+1}^{N} \partial c_j / \partial x = 0$ allow (12) to be rewritten

$$J_i = \sum_{j=1}^{N-1} D_{ij} \frac{\partial c_j}{\partial x} \ (j = 1, 2 \dots N\text{-}1) \qquad \cdot \qquad \cdot \quad (13)$$

---

† D_A^* and D_B^* are sometimes referred to as the *self-diffusion coefficients of the alloy*. This is a perfectly acceptable alternative terminology. But there is a tendency nowadays to employ the term 'self' to the extent of describing tracer impurity diffusion coefficients (*v.i*) as impurity self-diffusion coefficients. This latter term is ambiguous and misleading and its use is to be discouraged.

so that $(N-1)^2$ coefficients suffice to describe the diffusion behaviour. The analogue of Fick's second law is

$$\frac{\partial c_i}{\partial t} = \sum_{j=1}^{N-1} \frac{\partial}{\partial x}\left(D_{ij}\frac{\partial c_j}{\partial x}\right) \qquad . \qquad . \qquad . \qquad (14)$$

These equations have been applied to a few ternary systems (Table 4a).

It is possible to show from the principles of irreversible thermodynamics that not all the D_{ij} are independent and that a total of only $N(N-1)/2$ coefficients are in fact sufficient to describe diffusion in an N-component system. No measurements in metals have employed this reduced scheme of coefficients, for to do so requires a knowledge of the thermodynamic properties of the system that is rarely available.

METHODS OF MEASURING D

STEADY STATE METHODS

These are based directly on Fick's first law. The usual procedure is to maintain concentrations of diffusant on the opposite sides of a sample, which is usually a thin sheet or a thin-walled tube, and to measure the resulting steady rate of flow J. This is generally practicable only when the diffusing element is a gas or can be supplied to and removed from the sample through a vapour phase. If the surface concentrations c_1 and c_2 in equilibrium with the ambient atmospheres are known, an average D over this concentration range is, for a sheet of thickness t for example, simply $\bar{D} = Jt/(c_1 - c_2)$ (Method Ib). Alternatively, if the steady concentration distribution across the sample is determined $D(c)$ may be calculated from $D = J(\partial c/\partial x)$ (Method Ia).

D may also be calculated from measurements of the time required to reach a steady state (Method Ic).

These methods are used for measuring D only for interstitial solute diffusion: the Kirkendall Effect complicates any attempt to apply it reliably to substitutional diffusion.

NON-STEADY STATE METHODS

The change in the concentration distribution in a sample as a result of diffusion is measured and D deduced from a solution of Fick's second law [equations (2), (4), (5) or (14)] appropriate to the conditions of the experiment. There are three common types of experimental arrangement, two of which are usually employed in chemical diffusion coefficient measurements, the third in measurements of tracer diffusion coefficients.

(i) Diffusion Couple Method

Two metals, or two different homogeneous alloys of concentrations c_1 and c_2, are brought into intimate contact across a plane interface, say by welding. Diffusion is allowed to take place by annealing at a constant temperature for a time t. The distribution of concentration in the sample is then determined in some convenient manner, often by removal and subsequent analysis of a succession of thin layers cut parallel to the initial interface. It is usually arranged that the two halves of the couple be sufficiently thick that the diffusion zone does not extend to either end.

D generally varies with concentration, but no analytic solutions of (4) are available so recourse is had to a graphical method of analysis known as the Matano–Boltzmann method. The concentration c is plotted against x and $D(c)$ determined graphically from

$$D(c) = (2t \cdot \partial c/\partial x)^{-1}\int_{c}^{c_1} x\,\mathrm{d}c \quad \text{[Method IIa(i)]} \qquad . \qquad . \qquad (15)$$

The origin of x is located by the condition $\int_{c_1}^{c_2} x\,\mathrm{d}c = 0$ and this may be shown to coincide, under ideal conditions, with the *initial* position of the interface between the two members of the couple. Thus it is $\hat{D}$ which is measured in substitutional diffusion. Markers inserted at the interface locate its *final* position after diffusion.

It has already been mentioned that measuring their displacement x_m from $x = 0$ allows the partial diffusion coefficients to be calculated.

If D varies little in the range c_1 to c_2, and this is often so if the range is sufficiently restricted, equation (5) may be used, the solution of which for this case is

$$\frac{c - c_2}{c_1 - c_2} = \tfrac{1}{2}\left\{ 1 - \text{erf}\left(\frac{x}{2\sqrt{Dt}}\right)\right\} \qquad . \qquad . \qquad . \quad (16)$$

With $x = 0$ defined as before, D can then be calculated directly by a ' least squares ' fit of the $c \sim x$ data to this or other appropriate equations [Method IIa(ii)].

Occasionally, the diffusion couple method is used to measure self-diffusion coefficients, one half the couple being normal metal, the other enriched in one of its active or normal isotopes. It may be used too to measure diffusion coefficients in liquids (Shear-cell method).

With analytic solutions, like (16), D can be calculated by measuring c at one position only. This is sometimes done but it is not to be expected that values derived in this way will be as reliable as when derived from a complete $c \sim x$ curve (Method IIb).

The concentration range in a diffusion couple may span any number of phase regions in the equilibrium diagram of the system: the diffusion zone then consists of phase layers with concentration discontinuities across each boundary between two layers. In such cases equation (15) [Method IIa(i)] is still applicable. If D is assumed constant analytic solutions are available and with these it is sometimes possible (Method IIc) to determine D from measurements only of the rates of movement of one or more phase boundaries and knowledge of the equilibrium concentrations at the boundaries.[5]

(ii) *In-diffusion and Out-diffusion Methods*

Material is allowed to diffuse into, or out of, an initially homogeneous sample of concentration c_1 under the condition that the concentration at the surface is maintained at a constant and known value c_0 by being exposed to a constant ambient atmosphere. c_1 is usually zero for in-diffusion experiments and so is c_0 for out-diffusion experiments.

D may be calculated from a measurement either of the total amount of material taken up by or lost from the sample (Method IIIb), or of the concentration distribution within the sample after diffusion (Method IIIa). The first method gives an average D over the range c_1 to c_0. For the second, equation (15) can be used again to give $D(c)$ or, if D is constant, it may be calculated from an appropriate analytic solution.

When the loss (or gain) of material from the sample entails the movement of a phase boundary D can again be calculated from the rate of movement (Method IIIc). This method has been mostly used for interstitial solute diffusion, but also occasionally for substitutional diffusion measurements in systems with a sufficiently volatile component. A disadvantage of it is that conditions at the surface may not always be under adequate control so that c_0 is either ill-defined or not constant or both, with consequent uncertainty in D.

A common method of measuring liquid self-diffusion rates employs a type of out-diffusion method. A capillary tube, closed at one end and containing activated material, is immersed open-end uppermost in a large bath of inactive material. After the diffusion anneal the depleted activity content of the capillary is determined, and D calculated on the assumption that diffusion of the active species out of the tube is subject to zero concentration being maintained at the exit.

For determining the concentration distribution $c(x)$ in any of the above chemical diffusion methods a wide variety of techniques has been employed, including chemical and spectrographic analysis, X-ray and electron diffraction, electron microprobe analysis, X-ray absorption, microhardness measurements and so forth. In this edition of the tables the method of analysis is not recorded for it is probably of less importance in assessing the reliability of a result than other features of the experimental procedure.

(iii) *Thin Layer Methods*

These are used now almost exclusively for the measurement of self and of tracer D's. A very thin layer of radioactive diffusant, of total amount g per unit area, is deposited on a plane surface of the sample, usually by evaporation or electrodeposition.   After diffusion for time t the concentration at a distance x from the surface is

$$c(x) = \frac{g}{(\pi Dt)^{\frac{1}{2}}} \exp\left(-\frac{x^2}{4Dt}\right) \qquad . \qquad . \qquad . \quad (17)$$

provided the layer thickness is very much less than $(Dt)^{\frac{1}{2}}$.   This condition is easy to satisfy because extremely small quantities suffice for studying the diffusion on account of the very high sensitivity of methods of detecting and measuring radio-active substances.   For the same reason there is a negligible change in the chemical composition of the sample so D is constant and equation (5), of which (17) is the solution for this case, is applicable.

After diffusion the activity of each of a series of slices cut from the sample may be determined and D calculated from the slope $(=1/4Dt)$ of the linear plot of log activity in each slice against x^2 [Method IVa(i)].   Alternatively, such a plot may be constructed from intensity measurements made on an autoradiograph of a single section cut along or obliquely to the diffusion direction [Method IVa(ii)].

Another method is to calculate D from measurements made, after the removal of each slice, of the residual activity emanating from each newly exposed surface *of the sample* (Residual activity Method IVb).

Or, D may be determined by comparing the total activity from the surface $x = 0$ after diffusion with the original activity at $t = 0$ (Surface decrease Method IVc).

Methods IVb and c require an integration of equation (17).   They are generally regarded as less reliable in principle than Method IVa because they obviously necessitate also a knowledge of the absorption characteristics of the radiation concerned. In addition Method IVc is particularly susceptible to errors arising from possible oxidation and from evaporation losses of the deposited material.

INDIRECT METHODS, NOT BASED ON FICK'S LAWS

In addition to macroscopic diffusion there are a number of other phenomena in solids which depend for their occurrence on the thermally activated motion of atoms. From suitable measurements made on some of these phenomena it is possible to determine a D.   The more important of these are:

(1) Internal friction due to a stress-induced redistribution of atoms in interstitial solution in metals (Snoek effect). (Method Va.)

(2) A similar phenomenon occurring in substitutional solid solution and due, it is believed, to stress-induced changes in short range order (Zener effect).   (Method Va.)

(3) Phenomena associated with nuclear magnetic resonance absorption, especially the 'diffusional narrowing' of resonance lines and a contribution, arising from atomic mobility, to the spin-lattice relaxation time T_1 (Method Vb).

(4) Some magnetic relaxation phenomena in ferromagnetic substances (Method Vc).

(5) The sintering of metal powder particles or wires (Method Vd).

(1), (2), (3) and (4) are associated with atomic motion over only a few atomic distances, and so have the advantage of providing measurements of D at temperatures lower than are often practicable by conventional methods.   Since in every case measurements are made in homogeneous material the diffusion coefficients obtained are of the nature of tracer rather than chemical diffusion coefficients.

Most measurements of D are conducted at a series of temperatures so as to provide values of the constants A and Q occurring in the Arrhenius equation

$$D = A \exp(-Q/RT) \qquad . \qquad . \qquad . \quad (18)$$

which usually describes very well the observed temperature dependence.* A is called the 'frequency factor' and Q the activation energy.   Wherever possible, experimental measurements are reported in the tables in terms of A and Q alone.

Experiments may be made by any of the above methods either with single crystal or polycrystalline material.   With polycrystals there is, in addition to diffusion through the grains (volume diffusion), diffusion at a more rapid rate locally through the disordered regions of grain boundaries.   This can, however, be reduced to a negligible proportion of the whole by using large grain material and by working at relatively high temperatures because, since $Q_{gb} < Q_v$, grain boundary diffusion rates

---

* This is often true even of $\tilde{D}$, because Q_A and Q_B for the partial diffusion coefficients do not seem to differ very much.

increase less rapidly with temperature than do volume diffusion rates. Obviously single crystals are to be preferred in accurate measurements of what is intended to be volume diffusion but even in their case there may be, at too low temperatures, a contribution to D from diffusion along dislocations. Measured values of D will then tend to be above the values expected from an extrapolation of the high temperature data using (18), and when they do so to a noticeable extent are often discarded in estimating Q and A.

From measurements of the concentration distribution around a grain boundary— usually in a bicrystal into which material diffuses parallel to the boundary—a product $D'\delta$ may be deduced.[12] D' is the coefficient for diffusion *in* the boundary of width δ. δ is an uncertain quantity but all results quoted in Table 5 give values for A_{gb} calculated assuming $\delta = 5 \cdot 0 \times 10^{-8}$ cm. $D'\delta$ is found to depend on the orientation of the boundary and on the direction of diffusion within it.

MECHANISMS OF DIFFUSION

Most theoretical discussions of diffusion are concerned with an understanding of A and Q rather than of D itself. On the basis of theoretical calculations of Q for various possible mechanisms of diffusion and comparison with observed values, it has been supposed for some time that in metals atoms diffuse substitutionally by thermally activated jumps into vacant lattice sites—*i.e.* by the ' vacancy mechanism '. This has comparatively recently been very convincingly confirmed, at least for f.c.c. metals, by thermal expansion and quenching experiments. While the same mechanism is usually thought to operate in all other metal structures, there is some doubt at present whether this is in fact true for the b.c.c. metals βTi, βZr and γU—or at least whether the vacancy mechanism is the only one operating in their case.

For interstitial solutes there has never been any doubt that they diffuse by jumps from one interstitial position to another.

REFERENCES

Textbooks

1. P. G. Shewmon, " Diffusion in Solids," **1963**, New York: McGraw-Hill.
2. W. Seith and T. Heumann, " Diffusion in Metallen," 2nd edn. **1955**, Berlin: Springer-Verlag.
3. K.Hauffe, " Reaktionen in und an festen Stoffe," **1955**, Berlin: Springer-Verlag.
4. J. Crank, " The Mathematics of Diffusion," **1956**, Oxford: Clarendon Press.
5. W. Jost, " Diffusion in Solids, Liquids and Gases," 2nd edn., **1964**, New York: Academic Press.

Reviews

6. A. D. Le Claire, *Progr. Metal Phys.*, 1949, **1**, 306; 1953, **4**, 265.
7. C. E. Birchenall, *Metall. Rev.*, 1958, **3**, 235.
8. R. E. Howard and A. B. Lidiard, *Rep. Progr. Phys.*, 1964, **XXVII**, 246.
9. C. Tomizuka, " Methods of Experimental Physics " (edited by K. Lark-Horowitz and V. A. Johnson), **1959**, 6A, 364, New York: Academic Press.
10. D. Lazarus, *Solid St. Phys.*, 1960, **10**.

Papers

11. M. Cohen, C. Wagner and J. E. Reynolds, *Trans. A.I.M.E.*, 1953, **197**, 1534.
12. A. D. Le Claire, *Br. J. Appl. Phys.*, 1963, **14**, 351.

SUMMARY OF METHODS FOR MEASURING D

I. STEADY STATE METHOD with

    (*a*) Measurement of concentration distribution within the sample or,   Ia
    (*b*) Average gradient calculated from c_1 and c_2 as deduced from equilibrium data   Ib
    (*c*) Time-delay method (measurement of time to reach steady state)   Ic

NON-STEADY METHODS

II. *Diffusion Couple Methods*

    (*a*) With determination of $c \sim x$ curve and

        (i) Use of Matano Boltzmann analysis to give $D(c)$ IIa (i)
        (ii) When it is evident (or assumed) that D is effectively constant, calculation of D from an analytic solution   IIa (ii)
        (iii) When it is evident that D is *not* constant and an analytic solution is used to calculate a D corresponding to each value of c—giving an approximate $D(c)$ IIa (iii)

    (*b*) D calculated from a single concentration measurement   IIb
    (*c*) D calculated from an analytic solution, assuming D constant, using measurements of rate of movement of phase boundaries and knowledge of equilibrium concentrations on the boundaries   IIc

III. *In-diffusion and Out-diffusion Methods* In-(i) Out-(ii)

    (a) D calculated from $c \sim x$ curves   IIIa
    (b) D calculated from total gain or loss, or rate thereof   IIIb
    (c) D calculated from rate of phase boundary movement   IIIc

IV. *Thin Layer Methods*—with Radioactive Tracers

    (a) With measurement of activity $\sim x$ curve

        (i) By sectioning and counting   IVa (i)
        (ii) By autoradiography   IVa (ii)

    (b) Residual activity method   IVb
    (c) Surface decrease method   IVc

V. *Indirect Methods*

    (a) By internal friction   Va
    (b) By nuclear magnetic resonance   Vb
    (c) By ferro-magnetic relaxation   Vc
    (d) By sintering   Vd

NOTES ON THE TABLES

1. All measurements are reported whenever possible in terms of A and Q (see equation 18). A in cm²/sec: Q in kcal./mole: ($R = 1\cdot987$ cal/mole/°K 1 eV = 23 kcal./mole). Where errors are quoted these are authors' estimates.

2. The ' temperature range ' is the range over which measurements were used to calculate A and Q. Extrapolation too far outside this range may not in some cases give reliable values for D.

3. All alloy concentrations are in atomic percentages unless otherwise stated. Purity of material is as quoted and is presumably in weight percentages, although this is not always stated explicitly in papers.

4. s.c. = single crystals; p.c. = polycrystals.

5. In Table 4 a single concentration denotes the concentration at which $D(c)$ was determined. Two concentrations separated by a hyphen denote the range of concentration over which measurements were made. Where this is followed by a single D value, or a single set of A and Q values, it is also the concentration range over which these values are averages.

6. Bold type in Table 4. This is used: (1) To indicate the species to which the D's, or A and Q values, refer in cases where there might be ambiguity—usually for interstitial solid solutions. Where there is no bold type the data refer to the interdiffusion coefficients of the first two substitutional species. (2) To indicate which component was used in the vapour phase in experiments employing Methods I and III.

7. Where several measurements exist an attempt has been made to select what appear to be the most reliable one or two. Mostly these are later measurements and references to earlier work can usually be found by consulting the references quoted.

TABLE I.   SELF-DIFFUSION IN SOLID ELEMENTS

Group	Element	A	Q	Temp. range (°C)	Method	Ref.
IA	Li	$0\cdot24 {+0\cdot17 \atop -0\cdot10}$	$13\cdot2 \pm 0\cdot4$		Vb	1
		$0\cdot39 \pm 0\cdot02$	$13\cdot49 \pm 0\cdot07$	70–170	IIb Li⁶	2
	Na	$0\cdot242$	$10\cdot45 \pm 0\cdot03$	0–94	IIb Na²²	3
	K	$0\cdot31$	$9\cdot75$	0–60	IVa (i) K⁴² p.c.	4
	Rb	$0\cdot23$	$9\cdot4$	$-23-+40$	Vb	1
IB	Cu	$0\cdot20 \pm 0\cdot03$	$47\cdot12 \pm 0\cdot33$	685–1062 ⎱	IVa (i) Cu⁶⁴ s.c.	5
		$0\cdot327 (a)$	$48\cdot33$	765–1062 ⎰		
		$0\cdot33 {+0\cdot35 \atop -0\cdot17}$	$48\cdot2 \pm 1\cdot8$	860–1060	IVa (i) Cu⁶⁴ p.c.	6
	Ag	$0\cdot395 \pm 0\cdot035$	$44\cdot09 \pm 0\cdot18$	630–935 ⎱	IVa (i) Ag¹¹⁰ s.c. 99·99%	7
		$0\cdot49 (a)$	$44\cdot47$	715–935 ⎰		
	Au	$0\cdot091 \pm 0\cdot001$	$41\cdot70 \pm 0\cdot30$	704–1048	IVa (i) Au¹⁹⁸ p.c. 99·95%	8
IIA	Mg ∥ c	$1\cdot0$	$32\cdot2$ ⎱	468–635	IVa (i) Mg²⁸ s.c. 99·9+%	9
	⊥ c	$1\cdot5$	$32\cdot5$ ⎰			
IIB	Zn ∥ c	$0\cdot13$	$21\cdot8$ ⎱	240–410	IVa (i) Zn⁶⁵ s.c. 99·999%	10
	⊥ c	$0\cdot58$	$24\cdot3$ ⎰			
	∥ c	$0\cdot076$	$22\cdot0 \pm 0\cdot3$ ⎱	200–415	IVa (i) Zn⁶⁵ s.c. 99·99+%	11
	⊥ c	$0\cdot39$	$24\cdot9 \pm 0\cdot9$ ⎰			
	Cd ∥ c	$0\cdot05$	$18\cdot2 \pm 0\cdot3$ ⎱	130–310	IVa (i) Cd¹¹⁵ s.c. 99·5%	12
	⊥ c	$0\cdot10$	$19\cdot1 \pm 0\cdot7$ ⎰			
IIIB	Al	$1\cdot71$	$34\cdot0$	450–650	IVa (i) Al²⁶ p.c. 99·9%	13
	In ∥ c	$2\cdot7$	$18\cdot7 \pm 0\cdot3$ ⎱	44–144	IVa (i) In¹⁴⁴ s.c. 99·995%	14
	⊥ c	$3\cdot7$	$18\cdot7 \pm 0\cdot3$ ⎰			
	αTl ∥ c	$0\cdot4$	$22\cdot9 \pm 0\cdot5$ ⎱	150–225 ⎱	IVa (i) Tl²⁰⁴ s.c. 99·9+%	15
	⊥ c	$0\cdot4$	$22\cdot6 \pm 1\cdot0$ ⎰			
	βTl	$0\cdot7$	$20\cdot0 \pm 0\cdot5$	235–275 ⎰		
IVB	C (Graphite)	$0\cdot4 - 14\cdot4$	$163 \pm 12(f)$	2185–2347	IIIb C¹⁴ Natural Crystals	16
	Ge	$7\cdot8 \pm 3\cdot4$	$68\cdot50 \pm 0\cdot96$	766–928	IVa (i) Ge⁷¹ s.c.	17
		$10\cdot8 \pm 2\cdot4$	$69\cdot40 \pm 0\cdot44$	731–916	IVb Ge⁷¹ s.c.	18
	βSn ∥ c	$8\cdot2 \pm 6$	$25\cdot6 \pm 0\cdot8$ ⎱	178–222	IVa (i) Sn¹¹³ s.c. 99·998%	19
	⊥ c	$1\cdot4 \pm 0\cdot5$	$23\cdot3 \pm 0\cdot5$ ⎰			
	∥ c	$7\cdot7 \pm 3$	$25\cdot6 \pm 1\cdot0$ ⎱	160–228	IVa (i) Sn¹¹³ s.c. 99·999%	60
	⊥ c	$10\cdot7 \pm 1\cdot0$	$25\cdot1 \pm 0\cdot8$ ⎰			
	Pb	$0\cdot281$	$24\cdot21 \pm 0\cdot77$	174–322 ⎱	IVa(i) Pb²¹⁰ s.c. 99·999%	20
		$0\cdot69 (a)$	$25\cdot19 \pm 0\cdot17$	206–322 ⎰		
		$0\cdot46 {+0\cdot35 \atop -0\cdot20}$	$24\cdot80 \pm 0\cdot60$	200–300	IVa(i) Pb²¹⁰ p.c. 99·9999%	21

TABLE I. SELF DIFFUSION IN SOLID ELEMENTS—*continued*

Group	Element	A	Q	Temp. range (°C)	Method	Ref.
VB	αP	1.07×10^{-3} $(D = 1.07 \times 10^{-3} \exp(-9,400/RT) + 2.10^{46} \exp(-80,600/RT)$ for $30° < T <$ M.Pt.$)$	9.40 ± 0.225	0–30	IIb P^{32} p.c.	22
	Sb $\parallel c$	$22 \, {+22 \atop -11}$	47.1 ± 1.2	}	IVa(i) Sb^{124} s.c. 99.998%	23
	$\perp c$	$17 \, {+260 \atop -16}$	44.4 ± 4.9	} 490–610		
	Bi $\parallel c$	1.2×10^{-3}	31 (b) }	} 210–270		24
	$\perp c$	7.0×10^{46}	140 }			
VIB	S $\parallel c$	1.78×10^{36}	78.0 (g) }	} 37–94	IVa(i) S^{35} s.c.	55
	$\perp c$	8.32×10^{-12}	3.08 }			
	Se (Crystal)	$\{ \, {1.4 \times 10^{-4} \atop 7.6 \times 10^{-10}}$	$\{ \, {11.7 \atop 3.22 \, (h)}$	81–140 } 35–50 }	IVa(i) Se^{75} p.c.	56
	(Amorph.)	6.3×10^{25}	53	35–56 }		
IVA	αTi	6.4×10^{-8}	29.3	690–850	IVb Ti^{44} p.c. 99.89%	25
	βTi	$D = 3.58 \times 10^{-4} \exp(-31,200/RT) + 1.09 \exp(-60,000/RT)$ 900/1543° }			IVa(i) Ti^{44} p.c. 99.9% }	57
	αZr	5.9×10^{-2}	52.0	650–827	IVc Zr^{95} p.c. 99.6%	26
		5.6×10^{-4}	45.0	750–850		27
	βZr	$3.10^{-6}\left(\dfrac{T}{1136}\right)^{15.6}$	$19.6 + 0.031 \times (T - 1136)$	900–1750	IVa(i) Zr^{95} p.c. 99.94%	28
		0.04	44.8 (l)	900–1200	IVb Zr^{95} p.c. (Hf. free iodide)	61
	αTh	100–5,000	82.8 ± 7.0	1100–1400 }	Vd ' s.c.' ~99.9%	29
	βTh	$10^4 – 10^6$	99.0 ± 7.0	1450–1550 }		
VA	V	0.36 ± 0.02	73.65 ± 0.15	880–1356 }	IVa(i) V$^{48, 49}$ p.c. and s.c. 99.99%	30
		214.0 ± 20	94.14 ± 0.33	1356–1840 }		
	Nb	1.1 ± 0.2	96.0 ± 0.9	900–2400	IVa(i) Nb^{95} p.c.	31
	Ta	0.124	98.7	1200–2300	IVa Ta^{182} p.c. and s.c.	33
VIA	Cr	0.28 ± 0.06	73.2 ± 4.0	1200–1600	IVb Cr^{51} p.c. 99.99%	34
		$0.20 + 0.1$	73.7 ± 0.9	1028–1545	IVa(i) Cr^{51} p.c. 99.98%	59
	Mo	$0.1 \, {-0.05 \atop +0.1}$	92.2 ± 2.6 (j)	1850–2350	IVa(i) Mo^{99} s.c. 99.98%	35
		1.8	110.0	2155–2540	IVb Mo^{99} p.c. 99.97%	62
	W	0.7	121.0 (k)	2000–3228	IVa(i) W^{185} s.c. and p.c. 99.99%	36 and 37
	αU ∥ (100)	2.0×10^{-3} $D = 1.8 \times 10^{-14}$	40.0	580–650	IIb Enriched U^{234} p.c.	38
	∥ (010)	$D = 0.72 \times 10^{-14}$	}	} 640	IVc U^{233} p.c.	39
	∥ (001)	$D = 0.66 \times 10^{-14}$	}			
	∥ (100)	$D = 19.5 \times 10^{-14}$	}	} 625.5	IVa(i) U^{235} s.c.	40
	∥ (010)	$D = 10^{-14}$	}			
	∥ (001)	$D = 19.5 \times 10^{-14}$	}			
	β-U	1.35×10^{-2}	42.0	700–755	IIb Enriched U^{234} p.c.	41
	γ-U	1.8×10^{-3}	27.5	800–1040	IIb Enriched U^{234} p.c. 99.9%	42
		2.33×10^{-3}	28.5	803–1069	IVa(i) U^{235} p.c. 99.96%	43
VIII	α-Fe(Para)	1.9 ± 0.5	57.2 ± 1.1	809–905 }	IVb and IVc Fe^{55} p.c. and s.c.	44
	α-Fe(Ferro)	~2 (e)	~60	700–750 }		
	γ-Fe	0.18 ± 0.05	64.5 ± 0.65	1063–1393	IVc Fe^{55} p.c. 99.97%	
		0.22	64.0 ± 2.2	1156–1349 }	IVa(ii) Fe^{55} p.c. 99.998%	45
	δ-Fe	6.8	61.7 ± 4	1407–1515 }		
		1.9	57.0 ± 3	1413–1507	IVa(i) Fe^{59} p.c.	46
	Co	$0.35 \, {+0.33 \atop -0.17}$	65.0 ± 1.95	1015–1300	IVa(i) Co^{60} p.c. 99.4%	47
	Co(Ferro)	0.50 (e)	65.4	772–1048 }	IVb Co^{60} p.c. 99.2%	48
	Co(Para)	0.17	62.2	1192–1297 }		
	Ni	3.36	69.8	1150–1400	IVa(i) and IVb Ni^{63} p.c. 99.92%	50
		2.59 ± 0.45	69.5 ± 0.5	1085–1303	IVa(i) Ni^{63} p.c. 99.98%	51
	Pd	$0.205 \, {+0.05 \atop -0.04}$	63.6 ± 0.65	1050–1500	IVa(i) Pd^{103} s.c. 99.999%	52
	Pt	0.33	68.2 ± 1.4	1325–1600	IVa(i) Pt^{195m} p.c. 99.99%	53
		0.22 ± 0.03	66.6 ± 0.9	1250–1725	IVc Pt^{195m} p.c. 99.999%	54
	δ-Pu	4.5×10^{-3}	23.8	350–440	IIb Pu^{238} p.c.	58

Notes:

(a) Recalculated values ignoring measurements at the lower temperatures.

(b) Nachtrieb (unpublished work [23]) reports the anisotropy of Bi to be much less than given by these figures.

(c) The nature of the transition from the ferromagnetic to the paramagnetic region is shown in *Figure 1*.

(d) According to Ref. 45, the data for the α and δ regions may be represented by a single equation $D = 1.2 \exp(-55,800/RT)$. Ref. 45 also contains three measurements (at 863°, 880° and 899°C) which are in very good agreement with the more extensive measurements of Ref. 44.

(e) According to R. J. Borg [49] the measurements of Ref. 48 do not convincingly demonstrate a difference in the Q and A values in the paramagnetic and ferromagnetic regions. The data for these two regions, taken together, may be as well represented by the single set of values: $Q = 65.5$; $A = 0.51$.

(f) Assumed in analysis of results that diffusion in direction perpendicular to basal plane is negligible.

(g) Orthorhombic Sulphur crystals, but assumed in calculating results that $D_{11a} = D_{11b} = D\perp c$.
(h) Authors are of opinion that the higher values in the 35°–50°C range are due to grain boundary diffusion contributing appreciably.

(j) Consistently higher results were obtained with polycrystalline material, giving $A = 0\cdot5 \; {}^{+\;0\cdot7}_{-\;0\cdot3}$ and $Q = 96\cdot9 \pm 4\cdot4$.

(k) The values quoted represent the 'Best line' through the data points of (2000°–2700°) [36] and (2666°–3228°).[37] They are reported by Askill.[32]
(l) Values obtained from samples pre-annealed at 1200°C. Lower values of A and Q were reported from samples *not* pre-annealed.

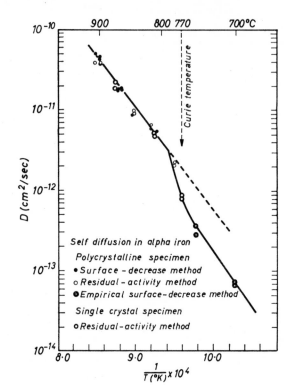

Figure 1.   Self-diffusion in αFe, above and below the Curie point[44]

REFERENCES TO TABLE 1

1. D. F. Holcomb and R. E. Norberg, *Phys. Rev.*, 1955, **98**, 1074.
2. A. N. Naumov and G. Ya. Ryskin, *Zhur. Tekh. Fiz.*, **29**, 189 and *Sov. Phys. Tech. Phys.*, 1959, **4**, 162.
3. N. H. Nachtrieb, E. Catalano and J. A. Weil, *J. chem. Phys.*, 1952, **20**, 1185.
4. J. N. Mundy, to be published.
5. A. Kuper, H. Letaw, L. Slifkin, E. Sonder and C. Tomizuka, *Phys. Rev.*, 1954, **96**, 1224; 1954, **98**, 1870.
6. K. Monma, H. Suto and H. Oikawa, *Nippon Kink. Gakk.*, 1964, **28**, 192.
7. C. T. Tomizuka and E. Sonder, *Phys. Rev.*, 1956, **103**, 1182.
8. S. M. Makin, A. H. Rowe and A. D. Le Claire, *Proc. Phys. Soc.*, 1957, **B70**, 545.
9. P. G. Shewmon, *J. Metals, N.Y.*, 1956, **8**, 918.
10. H. B. Huntington, G. A. Shirn and E. S. Wajda, *Acta metall.*, 1952, **1**, 513.
11. F. E. Jaumot and R. L. Smith, *J. Metals, N.Y.*, 1956, **8**, 137.
12. E. S. Wajda, G. A. Shirn and H. B. Huntington, *Acta metall.*, 1955, **3**, 39.
13. T. S. Lundy and J. F. Murdock, *J. appl. Phys.*, 1962, **33**, 1671.
14. J. E. Dickey, *Acta metall.*, 1959, **7**, 350.
15. G. A. Shirn, *Acta metall.*, 1955, **3**, 87.
16. M. A. Kanter, *Phys. Rev.*, 1957, **107**, 655.
17. H. Letaw, M. W. Portney and L. Slifkin, *Phys. Rev.*, 1956, **102**, 636.
18. H. Widmer and G. R. Gunther-Mohr, *Helv. phys. acta*, 1961, **34**, 635.
19. J. D. Meakin and E. Klokholm, *Trans. metall. Soc. A.I.M.E.*, 1960, **218**, 463.
20. N. H. Nachtrieb and G. S. Handler, *J. chem. Phys.*, 1955, **23**, 1569.
21. J. B. Hudson and R. E. Hoffman, *Trans. metall. Soc. A.I.M.E.*, 1961, **221**, 761.
22. N. H. Nachtrieb and G. S. Handler, *J. chem. Phys.*, 1955, **23**, 1187.
23. H. B. Huntington, P. B. Ghate and J. H. Rosolowski, *J. appl. Phys.*, 1964, **35**, 3027.
24. W. Seith, *Z. Electrochem.*, 1933, **39**, 538.
25. C. M. Libanati and Sra. F. Dyment, *Acta metall.*, 1963, **11**, 1263.
26. V. S. Lyashenko, V. N. Bykov and L. V. Pavlinov, *Fizika. Metall.*, 1959, **8**, 362.
27. P. Flubacher, *Rep. EIR. Bericht* No. 49 (Switzerland), 1963.
28. J. I. Federer and T. S. Lundy, *Trans. metall. Soc. A.I.M.E.*, 1963, **227**, 592.

29. C. J. Meechan, 2nd Nucl. Engng. Sci. Conf. (Phil. Pa.) Pap. No. 57 NESC-7, *Amer. Soc. mech. Engrs.*, 1957.
30. R. F. Peart, "Diffusion in B.C.C. Metals." 1965 *A.S.M.* and *J. Phys. chem. Solids*, 1965, **26**, 1853.
31. T. S. Lundy, F. Winslow, R. E. Pawel and C. J. McHargue, *Trans. metall. Soc. A.I.M.E.*, 1965, **233**, 1533.
32. J. Askill, *Thesis*, Reading University, 1964.
33. R. E. Pawel and T. S. Lundy, *J. Phys. chem. Solids*, 1965, **26**, 937.
34. W. C. Hagel, *Trans. metall. Soc. A.I.M.E.*, 1962, **224**, 430.
35. J. Askill and D. H. Tomlin, *Phil. Mag.*, 1963, **8**, 997.
36. W. Danneberg, *Metall.*, 1961, **15**, 977.
37. R. L. Andelin, J. D. Knight and M. Kahn, *Trans. metall. Soc. A.I.M.E.*, 1965, **233**, 19.
38. Y. Adda and A. Kirianenko, *J. nucl. Mater.*, 1962, **6**, 130.
39. R. Resnick and L. L. Seigle, *J. nucl. Mater.*, 1962, **5**, 5.
40. S. J. Rothman, J. J. Hines, J. Gray and A. L. Harkness, *J. appl. Phys.*, 1962, **33**, 2113.
41. Y. Adda, A. Kirianenko and C. Mairy, *J. nucl. Mater.*, 1959, **1**, 300.
42. Y. Adda and A. Kirianenko, *J. nucl. Mater.*, 1959, **1**, 120.
43. S. J. Rothman, L. T. Lloyd, R. Weil and A. L. Harkness, *Trans. metall. Soc. A.I.M.E.*, 1960, **218**, 605.
44. F. S. Buffington, K. Hirano and M. Cohen, *Acta. metall.*, 1961, **9**, 434.
45. D. Graham and D. H. Tomlin, *Phil. Mag.*, 1963, **8**, 1581.
46. R. J. Borg, D. Y. F. Lai and O. H. Krikorian, *Acta metall.*, 1963, **11**, 867.
47. W. Lange, A. Hässner and K. Sieber, *Isotopentechnik*, 1962, **2**, 42.
48. K. Hirano, R. P. Agarwala, B. L. Averbach and M. Cohen, *J. appl. Phys.*, 1962, **33**, 3049.
49. R. J. Borg, *J. appl. Phys.*, 1963, **34**, 1562.
50. J. R. MacEwan, J. U. MacEwan and L. Yaffe, *Can. J. Chemistry*, 1959, **37**, 1623.
51. A. Ya. Shinyayev, *Fizika Metall.*, 1963, **15**, 100.
52. N. L. Peterson, *Phys. Rev.*, 1964, **136**, 568.
53. G. V. Kidson and R. Ross, *Proc. 1st UNESCO Conf. 'Radioisotopes in Sc. Res.'* 1958, **1**, 185, Oxford: Pergamon Press.
54. F. Catteneo, E. Germagnoli and F. Grasso, *Phil. Mag.*, 1962, **7**, 1373.
55. R. B. Cuddebach and H. G. Drickamer, *J. chem. Phys.*, 1951, **19**, 790.
56. B. I. Boltaks and B. T. Plachenov, *J. Tech. Phys. U.S.S.R.*, 1957, **27**, 2229.
57. J. F. Murdock, T. S. Lundy and E. E. Stansbury, *Acta metall.*, 1964, **12**, 1033.
58. R. E. Tate, E. M. Cramer and A. S. Goldman, *Trans. metall. Soc. A.I.M.E.*, 1964, **230**, 639.
59. J. Askill and D. H. Tomlin, *Phil. Mag.*, 1965, **11**, 467.
60. C. Coston and N. H. Nachtrieb, *J. phys. Chem.*, 1964, **68**, 2219.
61. G. B. Federov, *Metall. chystykh metallov.*, 1963, **IV**, 34.
62. L. V. Pavlinov and V. N. Bykov, *Fizika Metall.*, 1964, **18**, 459.

TABLE 2. TRACER IMPURITY DIFFUSION COEFFICIENTS

In Cu

Element	A	Q	Temp. range (°C)	Method			Ref.
Zn^{65}	0.34 ± 0.04	45.6 ± 0.9	605–1049	IVa(i)	s.c.		1
Ga^{72}	0.55	45.9		IVa(i)	s.c.		2
As^{76}	0.12	42.0		IVa(i)	s.c.		2
Ni^{63}	3.8 ± 0.2	56.8 ± 0.1	695–1061	IVa(i)	s.c.	99.99%	3
	2.7 ± 0.35	56.5 ± 0.28	742–1076	IVa(i)	s.c.	99.998%	4
Co^{60}	5.7 ± 0.34	55.2 ± 1.1	700–950	IVa(i)	s.c.		5
	1.93 ± 0.06	54.1 ± 0.08	843–1076	IVa(i)	s.c.	99.998%	4
Fe^{59}	1.4 ± 0.28	51.8 ± 0.47	830–1074	IVa(i)	s.c.	99.998%	4
	1.01 ± 0.23	50.95 ± 0.46	716–1056	IVa(i)	s.c.	99.998%	6
Mn^{54}	$0.08 - 0.5$	$41 - 46(a)$	1069	IVa(i)	s.c.	99.998%	4
Ag^{110}	0.63	46.5		IVa(i)	s.c.		2
Cd^{115}	0.935 ± 0.27	45.7 ± 0.9	725–950	IVa(i)	s.c.	99.98%	7
Sb^{124}	0.34 ± 0.12	42.0 ± 0.7	600–1002	IVa(i)	s.c.	99.99%	8
Pd^{103}	$1.71 \begin{smallmatrix} +0.23 \\ -0.21 \end{smallmatrix}$	54.37 ± 0.3	807–1056	IVa(i)	s.c.	99.999%	9
Au^{198}	0.69	49.7		IVa(i)	s.c.		2
Hg^{203}	0.35	44.0		IVa(i)	s.c.		2
Tl	0.71	43.3	785–996	IVa(i)	s.c.	99.999%	36

(a) Estimated values from a single measurement at 1069.2°. $D = 1.49 \times 10^{-8}$ cm²/sec.

In Ag

Element	A	Q	Temp. range (°C)	Method			Ref.
Cd^{115}	0.44 ± 0.05	41.7 ± 0.21	592–937	IVa(i)	s.c.	99.99%	10
In^{114}	0.41 ± 0.04	40.63 ± 0.20	612–936	IVa(i)	s.c.	99.99%	10
Sn^{113}	0.25 ± 0.03	39.30 ± 0.20	592–937	IVa(i)	s.c.	99.99%	10
Sb^{124}	0.169 ± 0.003	38.32	468–942	IVa(i)	s.c.	99.99%	11
Pd^{103}	$9.57 \begin{smallmatrix} +1.63 \\ -1.37 \end{smallmatrix}$	56.75 ± 0.30	735–940	IVa(i)	s.c.	99.999%	9
$Ru^{103/106}$	180 ± 70	65.8	793–945	IVa(i)	s.c.	99.99%	12
Cu^{64}	1.23 ± 0.25	46.1 ± 0.9	716–945	IVa(i)	s.c.	99.99%	13
Zn^{65}	0.54 ± 0.05	41.7 ± 0.2	643–924	IVa(i)	s.c.	99.99%	14
Ge^{71}	0.084 ± 0.042	36.5 ± 1.5	670–850	IVa(i)	p.c.	—	15
Ni^{63}	21.9 ± 4.7	54.77 ± 0.50	748–951	IVa(i)	s.c.	99.99%	16
Co^{60}	104 ± 76	59.9 ± 1.2	745–943	IVa(i)	s.c.	99.99%	17
Fe^{59}	2.42 ± 0.23	49.04 ± 0.21	718–928	IVa(i)	s.c.	99.99%	6
Au^{198}	0.85 ± 0.09	48.28 ± 0.25	718–925	IVa(i)	s.c.	99.99%	18
Hg^{203}	0.079 ± 0.008	38.1 ± 0.2	653–948	IVa(i)	s.c.	99.99%	13
Tl^{204}	0.15 ± 0.08	37.9 ± 1.5	644–801	IVa(i)	p.c.	—	15
Pb^{210}	0.22	38.1 ± 2	700–825	IVa(i)	p.c.	—	19

TABLE 2.   TRACER IMPURITY DIFFUSION COEFFICIENTS—*continued*

In Au

Element	A	Q	Temp. range (°C)	Method			Ref.
Hg[203]	$0.116 \begin{smallmatrix} + & 0.13 \\ - & 0.06 \end{smallmatrix}$	37.38 ± 1.60	600–1027	IIIa	p.c.	99.994%	20
Pt[195]	$7.6 \begin{smallmatrix} + & 4.7 \\ - & 2.9 \end{smallmatrix}$	60.90 ± 1.2	900–1056	IVa(i)	p.c. and s.c.	99.98%	21
Ni[63]	0.034 ± 0.007	42.0 ± 0.4	702–988	IVb	p.c.	99.93%	22
Co[60]	0.068 ± 0.014	41.6 ± 0.4	702–948	IVb	p.c.	99.93%	22
Fe[59]	0.082 ± 0.016	41.6 ± 0.4	754–948	IVb	p.c.	99.93%	22
Ag[110]	0.072 ± 0.08	40.20 ± 0.25	699–1008	IVa(i)	s.c.	99.99%	18

In Be

Element	A	Q	Temp. range (°C)	Method			Ref.
Ag[110]	6.2 ± 1.6	46.1 ± 0.9	650–910	IVb	p.c.	99.85%	82
Ag[110] ∥ c / ⊥ c	0.41 ± 0.15 / 1.98 ± 0.7	39.1 ± 1.6 / 45.7 ± 1.8 }	656–897	IVb	s.c.	99.85%	82
Fe[59]	0.53 ± 0.2	51.8 ± 1.1	700–1076	IVb	p.c.	99.85%	82

In Zn

Element	A	Q	Temp. range (°C)	Method			Ref.
In[114] ∥ c / ⊥ c	0.062 ± 0.008 / 0.14 ± 0.02	19.1 ± 0.1 / 19.6 ± 0.1 }	171–395	IVa(i)	s.c.	99.999%	23
Ag[110] ∥ c / ⊥ c	0.32 ± 0.02 / 0.45 ± 0.07	26.0 ± 0.1 / 27.6 ± 0.2 }	271–413	IVa(i)	s.c.	99.999%	23
Au[195] ∥ c / ⊥ c	0.97 ± 0.22 / 0.29 ± 0.12	29.7 ± 0.3 / 29.7 ± 0.5	315–415 / 347–415 }	IVa(i)	s.c.	99.999%	24
Cd[115m] ∥ c / ⊥ c	0.114 ± 0.008 / 0.117 ± 0.003	20.54 ± 0.08 / 20.42 ± 0.03 }	224–416	IVa(i)	s.c.	99.999%	24

In Al

Element	A	Q	Temp. range (°C)	Method			Ref.
Zn[65]	1.1 ± 0.4	30.9 ± 0.6	405–654	IVa(i)	p.c.	99.99%	25
Ni[63]	2.9×10^{-8}	$15.7(a)$					
Co[60]	1.1×10^{-6}	$19.9(a)$ }	359–630	IVb	p.c.	99.99%	26
Fe[59]	4.1×10^{-9}	$13.9(a)$					
Cr[51]	3.01×10^{-7}	$15.4(a)$	250–605	IVb	p.c.	99.95%	27
Mn[54]	0.22	28.8	450–650	IVa(i)	p.c.	99.99%	28

(a) Values of A and Q for Ni, Co, Fe and Cr appear to be anomalously low. This has been attributed [26, 27] to the very low solubility of these elements in Al, but such an interpretation has been seriously questioned in Ref. 29.

In In

Element	A	Q	Temp. range (°C)	Method			Ref.
Tl	0.049	15.5 ± 0.78	49–155	IVa(i)	p.c.	99.99%	53

In Sn

Element	A	Q	Temp. range (°C)	Method			Ref.
In[114] ∥ c / ⊥ c	34.1 ± 6.5 / 12.2 ± 2.5	25.8 ± 0.5 / 25.6 ± 0.5 }	180–221	IVa(i)	s.c.	99.998%	30
Co[60]	5.5 ± 0.5	22.0 ± 2.0					
Zn[65]	$(9.8 \pm 1.0)10^{-4}$	7.8 ± 0.7 }	140–217	IVb	s.c. and p.c.	—	54

In Pb

Element	A	Q	Temp. range (°C)	Method			Ref.
Au[195]	4.1×10^{-3}	9.35 ± 0.07	94–325	IVa(i)	s.c.	99.999%	31
Tl[204]	0.511 ± 0.2	24.33 ± 0.44	206–323 }	IVa(i)	p.c.	99.99%	32
Bi[210]	$D = 2.66 \times 10^{-10}$ / $D + 1.006 \times 10^{-9}$		290.4 / 322.7 }				

In Te

Element	A	Q	Temp. range (°C)	Method			Ref.
Se[75]	2.6×10^{-2}	28.6	320–440 }	IVa(i)	p.c.	—	81
Hg[203]	3.4×10^{-5}	18.7	270–440 }				

TABLE 2. TRACER IMPURITY DIFFUSION COEFFICIENTS—*continued*

In Ni

Element	A	Q	Temp. range (°C)	Method			Ref.
Cu⁶⁴	0·57 +0·61/−0·29	61·7 ± 2·2	1050–1360	IVa(i)	p.c.	99·95%	35
	0·724	61·0	850–1050	IVb	p.c.	99·95%	45
Co⁶⁰	0·75 ± 0·25	64·7 ± 1·95	748–1192	IVb	p.c.	99·98%	33
Fe⁵⁹	0·008	50·5	950–1200	IVa(i)	p.c.	—	34
	0·16	59·0(a)	940–1120	IVb	p.c.	99·95%	48
Cr⁵¹	1·10 +0·9/−0·5	65·1 ± 1·9	1100–1270	IVa(i)	p.c.	99·95%	47
Mo⁹⁹	1·6 × 10⁻³	51·0	1100–1420	IVb	p.c.	—	46
W¹⁸⁵	2·0 +0·8/−0·6	71·5 ± 1·1	1100–1300	IVa(i)	p.c.	99·95%	59
	1·13	71·0	1100–1275	IVa(i)	—	99·9%	49
Au¹⁹⁸	2·0	65·0	900–1100	IVa(ii)	p.c.	99·986%	37
C¹⁴	0·1	33·0	600–900	IVb	p.c.	—	50
Sb¹²⁴	(1·8 ± 0·27)10⁻⁵	27·0 ± 0·4	1020–1220	—	—	99·97%	52

(a) The authors report small changes in A and Q with change in purity.

In Fe

Element	A	Q	Temp. range (°C)	Method			Ref.
Ni⁶³	In α–Fe 9·9(a)	61·9 ± 2·0	Paramag. 800–900	IVb	p.c.	99·999%	40
	1·3 ± 0·33	56·0 ± 1·1(b)	Paramag. 810–900	IVb	s.c. and p.c.		
	1·4	58·7(c)	Ferrom. 600–680	IVb and c	s.c.	99·97%	41
	In γ–Fe 0·77 ± 0·2	67·0 ± 0·7	930–1050	IVb	s.c.		
	6·92	77·6 ± 2·0	1152–1400	IVc	p.c.	99·91%	38
Co⁶⁰	In α–Fe 118(a)	68·3 ± 2·0	Paramag. 800–904	IVb	p.c.	99·999%	40
	9·5	62·3 ± 1·0	Paramag. 830–888	IVa(i)	p.c.	99·97%	55
	In γ–Fe 1·25 +0·81/−0·51	72·9 ± 1·6	1140–1340	IVa(i)	p.c.	99·98%	42
	In δ–Fe 5·5	61·2 ± 3·0	1396–1500	IVa(i) and b	p.c.	—	56
Cu	In γ–Fe 3·0	61·0	800–1200	IVc		~99%(d)	57
Au¹⁹⁵	In α–Fe 31·0(a)	62·4 ± 1·2	Paramag. 800–900	IVb	p.c.	99·999%	40
C¹⁴	In α–Fe 6·2 × 10⁻³	19·2	350–850	IVb	p.c.	99·93%	58
(e)	In γ–Fe 0·1	32·4	900–1060				

(a) At $T < 800°C$ D becomes increasingly less than would be calculated from these A and Q, due to the onset of ferromagnetism. See *Figure* 2(a) (Ni), 2(b) (Co), 2(c) (Au).
(b) Reference 40 criticizes the estimated error in this Q value and re-estimates it to be ±7 kcal./mol.
(c) The authors report a smooth transition in the values of D from about 800°C at the bottom end of the linear Arrhenius range in the paramagnetic region, to about 700°C at the top end of what they represent as a linear Arrhenius ferromagnetic range.
(d) Contains 0·63% Mn and 0·13% C among other impurities.
(e) Results have also been published for Cr and W diffusion in α and γ-Fe (43) but D appears nearly constant over the first 100° or more of the γ-range, so it is difficult to accept the measurements as trustworthy.

In Co

Element	A	Q	Temp. range (°C)	Method			Ref.
Ni⁶³	In Ferromag. Co(a) 0·34±0·10	64·3 ± 1·9	772–1048	IVb	p.c.	—	33
	In Paramag. Co(a) 0·10±0·03	60·2 ± 1·8	1192–1297				
	1·25	72·1 ± 1·4	1152–1400	IVc	p.c.	99·5%	38
Fe	0·21	62·7	1104–1303	IVa(i)	p.c.	99·9%	39

(a) In Reference 44 it is suggested there is no justification for this separation into a ferromagnetic and paramagnetic region. It is claimed that a single straight line would fit the experimental results just as well over the whole range 772°–1297°C.

TABLE 2.   TRACER IMPURITY DIFFUSION COEFFICIENTS—*continued*

In Cr

Element	A	Q	Temp. range (°C)	Method			Ref.
Fe^{55}	$0.47 \begin{smallmatrix}+1.63\\-0.36\end{smallmatrix}$	79.3 ± 4.8	1245–1413	IVb	p.c.		61
Mo^{99}	2.7×10^{-3}	58	1100–1420	IVb	p.c.		46

In V

Cr	$\left(9.54\begin{smallmatrix}+13.7\\-5.62\end{smallmatrix}\right)10^{-3}$	64.6 ± 2.3	960–1200	IVb	p.c.	99.8%	62

In Mo

Element	A	Q	Temp. range (°C)	Method			Ref.
Co^{60}	$18 \begin{smallmatrix}+20\\-9\end{smallmatrix}$	106.7 ± 3.8	1850–2330	(a)	IVa(i)	s.c. and p.c.	99.98%	66
Nb^{95}	$14 \begin{smallmatrix}+14\\-7\end{smallmatrix}$	108 ± 3.2	1900–2275				
W^{185}	3.18	112.9 ± 1.0	1700–2100	IVa(ii) d(i)	p.c.	—	67
Re^{186}	0.097	94.7	1700–2100	IVa(i)	p.c.	—	60

(a) Samples annealed *in vacuo*. D is reported to be lower when samples are annealed in argon [66, 68].
E.g. $D = 54 \exp(-118,000/RT)$

In Nb

Element	A	Q	Temp. range (°C)	Method			Ref.
Fe^{55}	$1.5 \begin{smallmatrix}+0.6\\-0.4\end{smallmatrix}$	77.7 ± 1.2	1390–2100	IVa(ii)	p.c.	99.74%	68
Co^{60}	$0.74 \begin{smallmatrix}+0.37\\-0.25\end{smallmatrix}$	70.5 ± 1.6	1550–2030				
Sn^{113}	$0.14 \begin{smallmatrix}+0.1\\-0.05\end{smallmatrix}$	78.9 ± 2.2	1850–2390	IVa	p.c.	99.85%	66

In Zr

Element	A	Q	Temp. range (°C)	Method			Ref.
Nb^{95}	In β-Zr $9.10^{-6}\left(\dfrac{T°}{1136}\right)^{18.1}$	$25.1 + 0.0355 \times (T° - 1136)$	880–1750	IVa(i)	p.c.	99.94%	69
Cr^{51}	In α-Zr 1.19×10^{-8}	18.0	700–850	IVb	p.c.	99.9%	70
	In β-Zr 3.85×10^{-2}	41.0	700–850				
Sn^{113}	In α-Zr 1.0×10^{-8}	22.0		IVb	p.c.	99–99.7%	71
	In β-Zr 5.0×10^{-3}	39.0					
U	In β-Zr 0.46	47.3					72
Ta^{182}	In α-Zr 100	70.0	700–800	IVb	p.c.	99.6%	73
Fe	In α-Zr 2.5×10^{-2}	48.0		IVb	p.c.	—	74
	In β-Zr 4.0×10^{-2}	30.0					

In W

Element	A	Q	Temp. range (°C)	Method			Ref.
$Re^{183/184}$	275 ± 110	162.8 ± 2.5	2666–3228	IVa(i)	s.c.	99.99%	75
Fe^{59}	11.5	66.0	930–1240	—	p.c.	—	76

In Ta

Element	A	Q	Temp. range (°C)	Method			Ref.
Nb^{95}	0.23	98.7	921–2484	IVa(i)	s.c.	99.7	77
Fe^{59}	0.505	71.4	930–1240	—	p.c.	—	76

TABLE 2. TRACER IMPURITY DIFFUSION COEFFICIENTS—*continued*

In U

Element	A	Q	Temp. range (°C)	Method			Ref.
In γ U							
Cu⁶⁴	$1.96 \begin{smallmatrix} +0.35 \\ -0.30 \end{smallmatrix} \times 10^{-3}$	24.06 ± 0.40	786–1040 ⎫				
Ni⁶³	$5.36 \begin{smallmatrix} +0.86 \\ -0.74 \end{smallmatrix} \times 10^{-4}$	15.66 ± 0.35	786–1040				
Co⁶⁰	$3.51 \begin{smallmatrix} +0.95 \\ -0.75 \end{smallmatrix} \times 10^{-4}$	12.57 ± 0.58	784–990				
Fe⁵⁹	$2.69 \begin{smallmatrix} +0.43 \\ -0.37 \end{smallmatrix} \times 10^{-4}$	12.01 ± 0.34	787–990 ⎬	IVa(i)	p.c.	99·99%	78
Mn⁵⁴	$1.81 \begin{smallmatrix} +1.95 \\ -0.94 \end{smallmatrix} \times 10^{-4}$	13.88 ± 1.66	787–939				
Cr⁵¹	$5.47 \begin{smallmatrix} +1.04 \\ -0.87 \end{smallmatrix} \times 10^{-3}$	24.46 ± 0.43	797–1038				
Nb⁹⁵	$4.87 \begin{smallmatrix} +1.18 \\ -0.94 \end{smallmatrix} \times 10^{-2}$	39.65 ± 0.50	790–1103 ⎭				
Au	$(4.86 \pm 1.3) \times 10^{-3}$	30.4 ± 0.520	784–1007	IVa(i)	p.c.	99·99%	79
In β U							
Cr⁵¹	$D = 1.77 \times 10^{-9}$		748·2 ⎫				
Fe⁵⁹	$D = 8.71 \times 10^{-9(a)}$		701 ⎬	IVa(i)	p.c.	99·993%	80
	$D = 2.60 \times 10^{-8(a)}$		760 ⎭				

(a) The mean of two values.

In Ti

	A_1	A_2	Q_1	Q_2				
		(a)		(a)				
In β Ti								
V⁴⁸	3·4	1.0×10^{-3}	61·5	34·7	900–1540	IVa(i)	p.c.	99·9%	63 and 65
Cr⁵¹	14·0	7.4×10^{-3}	65·5	36·6	970–1650	IVa(ii) ⎫			
Mn⁵⁴	12·0	7.6×10^{-3}	64·5	34·3	930–1650	IVa(i)			
Fe⁵⁵	15·0	8.0×10^{-3}	60·7	30·0	920–1650	IVa(ii) ⎬	p.c.	99·7/99·9%	64 and 65
Co⁶⁰	16·0	13.0×10^{-3}	61·3	30·9	910–1650	IVa(i)			
Ni⁶³	20·0	17.0×10^{-3}	60·0	31·6	930–1650	IVa(ii) ⎭			
Sc⁴⁶	—	2.1×10^{-3}	—	32·7	919–1290	IVa ⎫			
Sn¹¹³	9·5	0.38×10^{-3}	69·2	31·6	950–1600	IVa ⎬	p.c.	99·7/99·9%	65
Nb⁹⁵	9·5	1.3×10^{-3}	69·5	34·9	1000–1650	IVa(ii) ⎫	p.c.	99·7/99·9%	64 and 65
Mo⁹⁹	3·6	0.7×10^{-3}	65·0	36·9	900–1650	IVa(i) ⎭			
P³²	5·0	3.62×10^{-3}	56·5	24·1	945–1600	IVa	p.c.	99·7/99·9%	65

(a) The Arrhenius plots for diffusion in Ti show a distinct positive curvature and present results may best be represented by the equation:

$$D = A_1 \exp(-Q_1/RT) + A_2 \exp(-Q_2/RT)$$

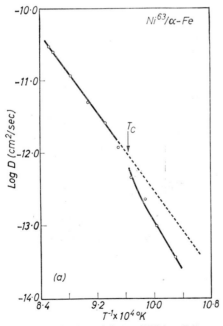

Figure 2(a). Tracer diffusion of Ni⁶³ into αFe⁴⁰

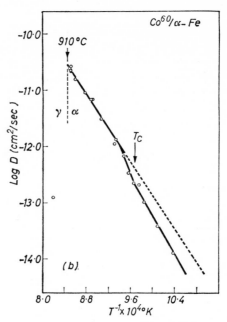

Figure 2(b).   Tracer diffusion of Co⁶⁰ into αFe⁴⁰

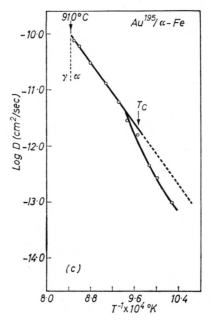

Figure 2(c).   Tracer diffusion of Au¹⁹⁵ into αFe⁴⁰

REFERENCES TO TABLE 2

1. J. Hino, C. Tomizuka and C. Wert, *Acta metall.*, 1957, **5**, 41.
2. C. Tomizuka (to be published). Quoted from D. Lazarus " Solid State Physics," 1960, Vol. 10.
3. A. Ikushima, *J. Phys. Soc., Japan*, 1959, **14**, 1636.
4. C. A. Mackleit, *Phys. Rev.*, 1958, **109**, 1964.
5. M. Sakamoto, *J. Phys. Soc., Japan*, 1958, **13**, 845.
6. J. G. Mullen, *Phys. Rev.*, 1961, **121**, 1649.
7. T. Hirone, N. Kanitomi, M. Sakamoto and H. Yamaki, *J. Phys. Soc. Japan*, 1958, **13**, 838.
8. M. C. Inman and L. W. Barr, *Acta metall.*, 1960, **8**, 112.
9. N. L. Peterson, *Phys. Rev.*, 1963, **132**, 2471.
10. C. T. Tomizuka and L. Slifkin, *ibid.*, 1954, **96**, 610.
11. E. Sonder, L. M. Slifkin and C. T. Tomizuka, *ibid.*, 1954, **93**, 970.
12. C. B. Pierce and D. Lazarus, *ibid.*, 1959, **114**, 686.
13. A. Sawatzky and F. E. Jaumot, Jr., *J. Metals*, 1957, **9**, 1207.
14. —— ——, *Phys. Rev.*, 1955, **100**, 1627.
15. R. E. Hoffmann, *Acta metall.*, 1958, **6**, 95.
16. T. Hirone, S. Miura and T. Suzuoka, *J. Phys. Soc., Japan*, 1961, **16**, 2456.
17. —— and H. Yamamoto, *ibid.*, 1961, **16**, 455.
18. W. C. Mallard, A. B. Gardner, R. F. Bass and L. M. Slifkin, *Phys. Rev.*, 1963, **129**, 617.
19. R. E. Hoffmann, D. Turnbull and E. W. Hart, *Acta metall.*, 1955, **3**, 417.
20. A. J. Mortlock and A. H. Rowe, *Phil. Mag.*, 1965, **11**, 1157.
21. —— —— and A. D. LeClaire, *ibid.*, 1960, **5**, 803.
22. D. N. Duhl, K. Hirano and M. Cohen, *Acta metall.*, 1963, **11**, 1.
23. J. H. Rosolowski, *Phys. Rev.*, 1961, **124**, 1828.
24. P. B. Ghate, *ibid.*, 1963, **131**, 174.
25. J. E. Hilliard, B. L. Averbach and M. Cohen, *Acta metall.*, 1959, **7**, 86.
26. K. Hirano, R. P. Agarwala and M. Cohen, *ibid.*, 1962, **10**, 857.
27. R. P. Agarwala, S. P. Muraka and M. S. Anand, *ibid.*, 1964, **12**, 871.
28. T. S. Lundy and J. F. Murdock, *J. appl. Phys.*, 1962, **33**, 1671.
29. R. W. Battoffi, *Acta metall.*, 1963, **11**, 1109.
30. A. Sawatsky, *J. appl. Phys.*, 1958, **29**, 1303.
31. A. Ascoli, *J. Inst. Metals.*, 1961, **89**, 218.
32. H. A. Resing and N. H. Nachtrieb, *Physics Chem. Solids*, 1961, **21**, 40.
33. K. Hirano, R. P. Agarwala, B. L. Averbach and M. Cohen, *J. appl. Phys.*, 1962, **33**, 3049.
34. M. B. Hyman and A. Ya. Shinyayev, *Dokl. Akad. Nauk*, SSSR, 1955, **102**, 969, and Aec. tr. 2262.
35. K. Monma, H. Suto and H. Oikawa, *Nippon Kink. Gakk.*, 1964, **28**, 192.
36. S. Komura and N. Kunitomi, *J. Phys. Soc. Japan*, 1963, **18** (Supp. II), 208.
37. A. D. Kurtz, B. L. Averbach and M. Cohen, *Acta metall.*, 1955, **3**, 442.
38. J. R. MacEwan, J. U. MacEwan and L. Yaffe, *Can. J. Chem.*, 1959, **37**, 1629.
39. H. W. Mead and C. E. Birchenall, *J. Metals*, 1955, **7**, 994.
40. R. J. Borg and D. Y. F. Lai, *Acta metall.*, 1963, **11**, 861.
41. K. Hirano, M. Cohen and B. L. Averbach, *ibid.*, 1961, **9**, 440.
42. T. Suzuoka, *Trans. Japan Inst. Metals*, 1961, **2**, 176.
43. P. L. Gruzin, *Dokl. Akad. Nauk. SSSR*, 1954, **94**, 681.
44. R. J. Borg, *J. appl. Phys.*, 1963, **34**, 1562.
45. S. P. Murarka, M. S. Anand and R. P. Agarwala, *J. appl. Phys.*, 1965, **36**, 3860.
46. P. L. Gruzin, S. V. Zemskii and I. B. Rodina, *Metall. Metallogr. Pure Metals*, Moscow, 1963, No. 4, p. 243. *A.E.R.E. Transl.* 1032.
47. K. Monma, H. Suto and H. Oikawa, *Nippon Kink. Gakk.*, 1964, **28**, 188.
48. P. Guiraldenq and P. Lacombe, " Colloque sur les Propriétés des Joints de Grains," 1960, p. 105, Saclay.
49. H. W. Allison and G. E. Moore, *J. appl. Phys.*, 1958, **29**, 842.
50. P. L. Gruzin, Yu. A. Polikarpov and G. B. Federov, *Physics Metals Metallogr*, 1957, **4**, 94.
51. ——.
52. P. P. Kuzmenko and G. P. Grinevich, *Soviet Phys. solid St.*, 1963, **4**, 2390.
53. R. E. Eckert and H. G. Drickamer, *J. chem. Phys.*, 1952, **20**, 13.
54. W. Chomka and J. Andruszkiewicz, *Nukleonika*, 1960, **5**, 611.
55. K. Sato, *Trans. Japan Inst. Metals*, 1964, **5**, 91.
56. R. J. Borg, D. Y. F. Lai and O. Krikorian, *Acta metall.*, 1963, **11**, 867.
57. R. Lindner and F. Karnik, *ibid.*, 1955, **3**, 297.
58. P. L. Gruzin, V. G. Kostogonov and P. A. Platanov, *Dokl. Akad. Nauk. SSSR*, 1955, **100**, 1069.
59. K. Monma, H. Suto and O. Oikawa, *Nippon Kink. Gakk.*, 1964, **28**, 197.
60. M. B. Bronfin, " Diffusion Processes, Structure and Properties of Metals " (Moscow, 1964). Translation, p. 24, 1965, New York: Consultants Bureau.
61. R. A. Wolf and H. W. Paxton, *Trans. metall. Soc. A.I.M.E.*, 1964, **230**, 1426.
62. R. A. Wolf, Thesis, *Carnegie Inst. Tech.*, 1962.
63. J. F. Murdock, T. S. Lundy and E. E. Stansbury, *Acta metall.*, 1964, **12**, 1033.
64. G. B. Gibbs, D. Graham and D. H. Tomlin, *Phil. Mag.*, 1963, **8**, 1269.
65. J. Askill and G. B. Gibbs, *Phys. Stat. Solid*, 1965, **11**, 557.
66. J. Askill, Thesis, Reading 1964, and *Proc. Gatlinburg Conf. Diffusion in B.C.C. Metals*, 1965. A.S.M.
67. S. Z. Bokshtein, M. B. Bronfin and S. T. Kishkin, " Diffusion Processes, Structure and Properties of Metals " (Moscow, 1964). Translation, p. 16, 1965, New York: Consultants Bureau.
68. R. F. Peart, D. Graham and D. H. Tomlin, *Acta metall.*, 1962, **10**, 519.
69. J. I. Federer and T. S. Lundy, *Trans. metall., Soc. A.I.M.E.*, 1963, **227**, 592.
70. R. P. Agarwala, S. P. Murarka and M. S. Anand, *Trans. metall. Soc. A.I.M.E.*, 1965, **233**, 986.
71. P. L. Gruzin, V. S. Emelyanov, G. G. Ryabora and G. B. Federov, *Geneva Conf. Proc.*, 1959, **19**, 187.
72. G. B. Federov and F. I. Zhomev, *Metallov. Met. Cristykh. Metal Sb. Nauchn. Rabot*, 1961, **3**, 193.
73. E. V. Borisov, V. G. Godin, P. L. Gruzin, A. I. Eustyukhin and V. S. Emelyanov, *Metallov. Met. Izdatel Akad. Nauk SSSR*, Moscow, 1959, and in Translation NP-TR-448.
74. A. M. Blinkin and V. V. Vorobiov, *Ukran. Fiz. Zh.* (U.S.S.R.), 1964, **9**, 91.
75. R. L. Andelin, J. D. Knight and M. Kahn, *Trans. metall. Soc. A.I.M.E.*, 1965, **233**, 19.
76. Y. P. Vasil'ev, I. F. Kamardin, V. I. Skatskii, S. G. Chermomorchenko and G. N. Schuppe, *Trudy. sred. -aziat. gos. Univ. Lenina*, 1955, **65**, 47.
77. R. E. Pawel and T. S. Lundy, *J. Phys. chem. Solids*, 1965, **26**, 937.
78. N. L. Peterson and S. J. Rothman, *Phys. Rev.*, 1964, **136A**, 842.
79. S. J. Rothman, *J. Nucl. Mater.*, 1961, **3**, 77.
80. ——, N. L. Peterson and S. A. Moore, *ibid.*, 1962, **7**, 212.
81. Sh. Merlanov and A. A. Kulier, *Soviet Phys. solid St.*, 1962, **4**, 394.
82. M. C. Naik, J. M. Dupouy and Y. Adda, Paper presented at Journée d'Automne 1964, *Société Français de Metallurgie*.

TABLE 3.　DIFFUSION IN HOMOGENEOUS ALLOYS

Element 1 (purity) at %	Element 2 (purity) at %	A_1^*	Q_1^*	A_2^*	Q_2^*	Temp. range (°C)	Ref.
Ag (—)	Al (—)		IVa(i)	p.c.			
	2·05	0·25	42·5	—	—		
	9·47	0·83	42·9	—	—	700–850	1
	14·1	0·73	41·2	—	—		
Ag (99·99) 100	Au (99·99) 0	0·49	44·47	s.c. 0·85	48·28		
	8	0·52	44·79	0·82	48·30		
	17	0·32	44·05	0·48	47·30		
	35	0·23	43·54	0·35	46·67		
	50	0·19	43·11	0·21	45·28	630–1010	2
	66	0·11	41·73	0·17	44·51		
	83	0·09	41·02	0·12	43·05		
	94	0·072	40·26	0·09	42·08		
0	100	0·072	40·2	0·09	41·7		
Ag (99·99) 100	Cd (99·999) 0	0·44	IVa(i) 44·27	s.c. and p.c. 0·44	41·69		
	6·50	0·31	42·61	0·33	40·48		
	13·60	0·23	40·96	0·22	38·61	500–900	3
	27·5	—	—	0·25	35·95		
	28·0	0·16	37·25	—	—		
(99·99)	(99·999)		IVa(i)	p.c.			
	33	0·11	36·19	0·24	34·95	530–770	63
	36	0·09	38·63	0·13	32·86		
Ag (—)	Cu (—)		IVa(i)	p.c.			
	1·75	0·66	44·8	—	—		
	4·16	1·84	46·6	—	—	700–890	1
	6·56	0·51	43·5	—	—		
Ag (—)	Ge (—)		IVa(i) and IVb	p.c.			
	1·50	0·55	44·0	D_{Ge} greater than for			
	3·00	1·59	45·3	infinite dilution by			1 and 7
	4·30	1·89	44·5	~15%/at. % Ge		700–850	
	5·43	2·18	44·2				
Ag (99·99) 100	In (99·99) 0	0·44	IVa(i) 44·27	s.c. and p.c. 0·41	40·80		
	4·40	0·36	42·67	—	—		
	4·70	—	—	0·45	40·30		
	12·40	—	—	0·57	38·39	500–900	3
	12·60	0·12	37·40	—	—		
	16·60	—	—	0·57	36·61		
	16·70	0·18	36·27	—	—		
Ag (99·95)	Mg (99·95)		IVb	p.c.			
	45·8	1·53	41·3	—	—		
	49·8	0·28	40·6	—	—	500–700	12
	52·0	0·134	38·0	—	—		
(99·98)	(99·9+)		IVa(i)	p.c.			
	41·10	0·095	33·2	—	—		
	43·60	0·15	35·3	—	—		
	48·48	0·37	39·5	—	—	500–700	13
	48·72	0·39	39·7	—	—		
	52·82	0·33	36·7	—	—		
	57·15	0·051	28·7	—	—	500–600	
	60·88	$D_{Ag}^* = 4\cdot37 \times 10^{-9}$ at 500·5°C					
Ag (—)	Pb (—)		IVa(i)	p.c.			
	0·21	0·22	42·5	—	—		
	0·25	—	—	0·22	37·8		
	0·52	—	—	0·38	38·7		
	0·71	0·89	44·7	—	—	700–800	9 and 1
	1·30	0·70	43·5	—	—		
	1·32	—	—	0·46	38·5		
Ag ('Spec. Pure')	Pd		IVa(i)	p.c.			
	0–21·8	$0\cdot27e^{-8\cdot2c}$	43·7	—(a)	—	715–942	4
	0–20·4	—(a)	—	$12\cdot5e^{-7\cdot5c}$	57·2	850–900	5

(a) $c = $ conc. of Pd.

TABLE 3. DIFFUSION IN HOMOGENEOUS ALLOYS—*continued*

Element 1 (purity) at %	Element 2 (purity) at %	A_1*	Q_1*	A_2*	Q_2*	Temp. range (°C)	Ref.
Ag (99·99)	Sb (99·99)		IVa(i)	s.c. (a)			
	0·53	0·38	43·5	—	—		
	0·89	0·30	42·6	—	—	} 550–900	6
	1·42	0·275	42·0	—	—		

(a) In 0·7 at % Sb alloys, D_{Sb}* same as D in pure Ag.
In 2·8 at % Sb alloys, D_{Sb}* ~20% greater than D in pure Ag.

Element 1 (purity) at %	Element 2 (purity) at %	A_1*	Q_1*	A_2*	Q_2*	Temp. range (°C)	Ref.
Ag (—)	Sn (—)			p.c.			
	0·18	0·13	41·7	—	—		} 24
	0·48	0·13	40·9	—	—		
	0·91	0·17	40·5	—	—		
	0·97	0·28	39·4	—	—	} 700–850	} 25
	2·8	0·1	38·6	—	—		
	4·56	0·23	39·7	—	—		} 24
	5·1	0·2	38·6	—	—		
	7·45	0·16	37·0	—	—		} 25
Ag (—)	Tl (—)		IVa(i) and IVb	p.c.			
100	0	0·724	45·5	0·15	37·9		
	1·1	0·42	43·5	0·72	40·4	} 640–800	7
	2·6	0·35	41·9	0·57	39·4		
	5·5	0·10	37·6	—	—		
Ag (99·99)	Zn (99·999)		IVa(i)	p.c.			
85	15		~36	0·11	36·0	500–700	10
70 (Merck.)	30	0·29	35·99	0·46	35·2		8
	(99·999)		IIb	p.c.			
52·4	47·6	$4·55 \times 10^{-3}$	17·6	—	—	400–610	11
Al (99·9)	Co (Carbonyl)		IVc	p.c.			
	10	—	—	2·65	67·5	1040–1220	14
42		—	—	$1·84 \times 10^2$	85	1000–1200	
49		—	—	333×10^2	102	1100–1300 }	15
50·7		—	—	$0·013 \times 10^2$	65	1000–1200	
	49/57		Individual D_{Co}* values plotted at 1250				16
Al (—)	Fe (—)		IVa	p.c.			
			Diff. of Co				
3·47		0·1	53·0	3·2	59·0		
7·95		1·9	56·0	4·5	60·0		
13·5		6·8	58·5	0·4	52·0		
20·6		22·0	60·0	32·0	63·0		
23·6		27·0	60·0	27·0	62·5	} —	17
35·5		210·0	67·0	—	—		
42·0		580·0	71·0	—	—		
47·3		6300·0	79·0	—	—		
52·0		148·0	67·0	60·0	66·0		
Al (—)	Ni (—)		IVc	p.c.			
			Diff. of Co⁶⁰				
	47·3	$4·7 \times 10^{-2}$	56·6	—	—		
	48·5	$9·3 \times 10^{-2}$	59·9	—	—		
	49·4	$4·4 \times 10^{-3}$	52·5	—	—	} 1050–1350	18
	50·7	57·7	80·6	—	—		
	53·1	2·6	67·6	—	—		
	55·1	$7·2 \times 10^{-3}$	47·1	—	—		
Al (99·99)	Zn (99·999)		IVa(i)	p.c.			
100	0	—	—	1·1	30·9	400–650	
	4·33	—	—	0·35	28·3	360–610	
	9·23	—	—	0·20	27·0	360–575	
	16·7	—	—	0·10	25·0	360–525	
	36·9	—	—	0·16	24·1	360–450 }	19
	49·4	—	—	0·048	21·9	325–440	
	62·9	—	—	0·12	20·0		
0	100·0	—	—	0·031	20·5	} 325–405	

TABLE 3. DIFFUSION IN HOMOGENEOUS ALLOYS—*continued*

Element 1 (purity) at %	Element 2 (purity) at %		$A_1{}^*$	$Q_1{}^*$	$A_2{}^*$	$Q_2{}^*$	Temp. range (°C)	Ref.
Au (99·99) 50 (99·999+)	Cd (99·95) 50 (99·999+)		0·17	IVa(i) 27·9	s.c. 0·23	28·0	300–590(a)	20
	47·5		0·23	IVa(i) 28·1	s.c. 1·36	31·0	370–570	
	49·0		0·61	30·0	1·5	31·2	440–550 }	21
	50·5		0·12	27·0	0·78	29·2	440–550	

(a) Between 590° and the m. pt. at 626° there is marked upward curvature in the Arrhenius plot.

Element 1 (purity) at %	Element 2 (purity) at %		$A_1{}^*$	$Q_1{}^*$	$A_2{}^*$	$Q_2{}^*$	Temp. range (°C)	Ref.
Au (99·96) 100	Ni (99·99) 0		0·26	IVa(ii) 45·3	p.c. 0·30	IVa(i) 46·0		
	10		—	—	0·80	47·6		
	20		0·05	40·2	0·82	47·8	Variable	
	35		0·063	42·7	—	—	(For Ni,	22 (Au)
	36		—	—	1·10	49·1	50–75°	
	50		0·091	43·4	0·09	44·5	temp.	23 (Ni)
	65		0·51	48·8	0·005	39·6	ranges	
	80		1·1	60·5	0·05	49·2	only)	
	90		—	—	0·04	51·1		
0	100		2·0	65·0	0·40	63·8		
C (—) V. small (—)	Fe (—) 100 (—)		0·008	Va and Vc 19·8(b) IVa(i)	p.c. —	—	50 to +150	26
0			—	—	0·44	67·0		
1·15			—	—	0·052	59·0		
2·46			—	—	0·015	54·0	1000–1300	28
3·39			—	—	0·021	54·0		
4·97			—	—	0·029	53·8		
6·21			—	—	0·050	53·8		
C (—) c% c%	Fe (—) (γ)	Ni (—) 20 25	—	IVb —	p.c. $18 \cdot 10^{-0.92c}$ $71 \cdot 10^{-0.65c}$	75–6c 79–5c	800–1300 } 1050–1330 }	29

(b) These values also represent the best fit to the original (Ref. 26) and later measurements when plotted altogether. See Ref. 27.

Element 1 (purity) at %	Element 2 (purity) at %		$A_1{}^*$	$Q_1{}^*$	$A_2{}^*$	$Q_2{}^*$	Temp. range (°C)	Ref.
C 0·104	Fe (99·97) (α)	Si 5·5	V(a) p.c. $D_e{}^*$ " virtually " the same as in Si-free Fe.				26–70	67
C V. small	Nb (99·4) ~100		0·0040	Va 33·02	p.c. —	—	130–280	30
C (—) V. small	Ta (—) ~100		0·0061	Va 38·51	p.c. —	—	190–360	30
C (—) V. small	V (—) ~100		0·0045	Va 27·29	p.c. —	—	60–165	30
Cd (—) 75 75 25 25	Mg (—) 25 25 75 75		$11 \cdot 2 \times 10^{-5}$ 0·074 $4 \cdot 10^{-5}$ $1 \cdot 2 \times 10^{-6}$	IVb 12·4(Ord.) 16·7(Disord.) 16·3(Ord.) 12·8(Disord.)	s.c. —	—	54–90 } 95–200 } 120–155 } 155–281 }	31 32
Co (—)	Cr (—) 4 7		0·67 56·3	IVb 65·8 79·3	p.c. —	— }	1100–1350	33
Co (—)	Cr (—) 9 18	Ni (—) 26 26	6·3 0·4	IVb 72·1 64·2	p.c. —	— }	1100–1350	33
Co (—) (98) (γ) (α)	Fe (—) (99) 21 50 50		0·54 $1 \cdot 1 \times 10^{-4}$ $2 \cdot 6 \times 10^{-6}$	IVb 65·0 IVa(i) 41·8 27·4	p.c. — p.c.	—	1100–1300 1000–1250 } 840–925 }	33 34
Co)	Fe ( ) 15	Ti ( ) 4	0·008	IVb 51·2	p.c. —	—	1100–1200	33

TABLE 3. DIFFUSION IN HOMOGENEOUS ALLOYS—*continued*

Element 1 (purity) at %	Element 2 (purity) at %	A_1*	Q_1*	A_2*	Q_2*	Temp. range (°C)	Ref.
Co (99·5) 100	Ni (99·98)		IVb	p.c.		Ferromag.(a)	
	0	0·5	65·4	0·34	64·3	772–1048	
	11	0·61	67·1	0·46	65·6	864–1048	
	20	5·96	73·5	1·66	69·7	845–1048	
	30	1·16	68·6	2·01	68·5	772–899	
	51	0·096	61·5	0·36	63·6	701–819	
100						Paramag.(a)	35
	0	0·17	62·2	0·10	60·2	1192–1297	
	11	0·21	63·6	0·17	62·7	1144–1297	
	20	2·42	71·0	0·41	65·7	1090–1246	
	30	0·78	67·0	0·67	64·8	1050–1250	
	51	0·12	60·2	0·21	60·6	900–1190	
	100	0·75	64·7	1·70	68·1	700–1190	
Co 5·25	(99·4%)	Diffusion of C^{14}		IVb p.c.			36
		0·4	37·0	—	—	600–900	

Co ( )	Mn ( )	Ni ( )	A_1*	Q_1*	A_2*	Q_2*	Temp. range (°C)	Ref.
19·5	20·3	60·0	0·86	IVb 60·5	p.c. —	—	1020–1120	
40·6	20·45	38·9	0·22	57·5	—	—	1020–1120	14
59·7	20·6	19·5	0·05	54·5	—	—	1060–1160	

(a) According to Ref. 64 measurements in pure Co, in Co + 11% Ni and Co + 20% Ni alloys do not convincingly demonstrate a difference in $A*$ and $Q*$ values for the paramagnetic and ferromagnetic regions.

Element 1	Element 2	A_1*	Q_1*	A_2*	Q_2*	Temp. range (°C)	Ref.
Cr	Fe (≥98)		IVa(ii) and IVb	p.c.			
12·5		1·29	55·1	—	—	} 850–1420	37
18·0		0·46	52·5	1·34	55·8		
26·0		0·156	48·4	—	—		
51·0		40·0	70·0	—	—	} 950–1320	38
69·0		24·6	75·5	—	—		
('Electrolytic')			IVa(i)	p.c.			
27·0		—	—	0·195	50·4	} 1040–1400	39
51·0		—	—	249·0	74·6		
84·0		0·376	64·2	146·0	81·9		
(b)Cr ('18–8' St. Steel with 0·7% Si, 0·1% Mo, 0·45% C)	Fe Ni		IVc	p.c.			41
		—	—	0·58	67·1	808–1200	
Cr (99·98)	N ( ) ~0·01		Va	p.c.			52
		—	—	$3\cdot10^{-4}$	24·3	50–170	
Cr	Ni (99·95)		IVa(i)	p.c.			
0		1·1	65·1	1·9	68·0	1100–1270 ($D_{Cr}*$)	
10		1·4	66·5	3·3	70·2	1040–1275 ($D_{Ni}*$)	42
20		1·9	67·7	1·6	68·5	1040–1404 (Ni Self D)	
35		IVa(i) p.c. 0·2	58·6	IVa(ii) $4\cdot10^{-3}$	49·0	890–1360	68

(b) Measurements for alloys between 20% and 80% Cr have also been reported in Ref. 40. There are inconsistencies among the figures presented which make it difficult to identify the true values. 'Recalculated' values from this paper are suggested in Ref. 37.

Cr	Ni (99·8)	Fe	p.c.	IVb	Diffusion of C^{14}		Temp. range (°C)	Ref.
0·74		0·52	—	—	0·1	34·0	600–900	36
4·65		0·36	—	—	0·5	37·0	500–900	

Cr (99·4)	Ti	p.c.	IVa(ii)	A_2*	Q_2*	Temp. range (°C)	Ref.
10·0		0·02	40·2	—	—	925–1180	44
18·0		0·09	44·5	—	—		

Cu (99·99)	Ni (99·95)	p.c.	IVa(i)	A_2*	Q_2*	Temp. range (°C)	Ref.
0		0·57	61·7	1·9	68·0	1050–1360	
13·0		1·5	63·0	35·0	74·9	1050–1360	
45·4		2·3	60·3	17·0	66·8	985–1210	43
78·5		1·9	55·3	0·063	49·7	860–1113	
100·0		0·33	48·2	1·7	55·3	860–1070	

TABLE 3. DIFFUSION IN HOMOGENEOUS ALLOYS—*continued*

Element 1 (purity) at %	Element 2 (purity) at %	$A_1{}^*$	$Q_1{}^*$	$A_2{}^*$	$Q_2{}^*$	Temp. range (°C)	Ref.
Cu ( )	Sb ( )	p.c.	IVc				
(δ)	19·4	$D_{Cu}{}^*$ 4·10^{-10}		$D_{Sb}{}^*$ 3·10^{-11}			
(χ)	24·4	7×10^{-9}		$3 \cdot 1 \times 10^{-10}$		390	61
(γ)	33·4	$\sim 10^{-9}$		$2 \cdot 7 \times 10^{-9}$			
Cu ('Spec. P') (α)	Zn ('Spec. P') 31	s.c. 0·34	IVa(i) 41·9	0·73	40·7	580–905	45
	(99·99)	s.c. 0·011	IVa(i) 22·04	0·0035	18·78	Disord. 497–817	
		180	37·09(a)	78·10^3	44·23	Ord. 380–450	
(β)	45·65 to 48·1	80	36·02(a)	163·0	36·3	Ord. 264–380	46
		Diffusion of Sb 0·08	23·5			Disord. 498–594	
(—)	(—)		p.c. IVa(i)				
(β)	46·7	0·020	23·42	0·022	22·04	Disord. 500–800	47
		0·80	31·0(b)	1·0	32·0	Ord. around 300	
(—)	(—)	p.c.	IVa(i)				
(β)	47·2	Diffusion of Ag 0·014	21·9(e)	Diffusion of Co 0·47	26·9	Disord. 470–700	48

(a) Arrhenius plots in the ordered region are curved. The values of A and Q reported describe straight line approximations to the data over the temperature ranges indicated.

(b) Ditto. The values of A and Q given here refer to the data at the lower end of the ordered temperature range investigated, viz: 300°C.

(e) Values of D* for the ordered region are shown only in graphical form in ref. 48.

Element 1 (purity) at %	Element 2 (purity) at %	$A_1{}^*$	$Q_1{}^*$	$A_2{}^*$	$Q_2{}^*$	Temp. range (°C)	Ref.
Cu (Electrolytic)	Zn (99·99)		IVb	s.c.			(d)
0		—	—	$\perp c$ 1·62	26		
0		—	—	$\parallel c$ 0·013	19		
0·2		—	—	$\perp c$ 3·2	26		
0·2		—	—	$\parallel c$ 0·021	19		
0·3		—	—	$\perp c$ 3·4	26	550–630	56
0·3		—	—	$\parallel c$ 0·025	19		
0·4		—	—	$\perp c$ 3·4	26		
0·4		—	—	$\parallel c$ 0·029	19		
0·5		—	—	$\perp c$ 3·5	26		
0·5		—	—	$\parallel c$ 0·035	20		
Fe (—)	Mn (—)		IVb	p.c.			
	0·4	0·2 × 10^{-3}	83·0	—	—		
	1·17	2·5 × 10^{-3}	91·0	—	—		
	2·28	8·0 × 10^{-3}	94·0	—	—		
	2·82	16·5 × 10^{-3}	96·0	—	—	950–1350	49
	4·39	24·0 × 10^{-3}	98·0	—	—		
	5·26	12·0 × 10^{-3}	95·0	—	—		
	6·15	4·0 × 10^{-3}	92·5	—	—		
	8·35	1·0 × 10^{-3}	89·0	—	—		
Fe ( ) ~100	N ( ) V. small	—	Va —	p.c. 0·005	18·4	−20—+160	51

(d) Additions of 0·5 at % Al to this range of Cu Zn alloys have a negligible effect on $Q_{Zn}{}^*$ but *decrease* $A_\perp$ by ~20% and $A_\parallel$ by ~40%.

Element 1 (purity) at %	Element 2 (purity) at %	$A_1{}^*$	$Q_1{}^*$	$A_2{}^*$	$Q_2{}^*$	Temp. range (°C)	Ref.
Fe (Electrolytic)	Ni (Electrolytic)		IVc	p.c.			
	5·8	—	—	2·11	73·5	1160–1390	50
	14·88	—	—	5·0	75·6		
(—)	(—)	—	IVb	p.c.		800–1300	29
	20·0	18	75	—	—	1050–1330	
	25·0	71	79	—	—		
Fe (Spec. P)	Ti (99·7)	p.c.	IVa(ii)	Diffusion of Nb[95]			
5		9·2 × 10^{-2}	39·6	1·82 × 10^{-3}	34·9	850–1260	
10		2·14	48·6	2·9 × 10^{-2}	41·3	850–1200	53
15		52·5	58·1	9·9	56·0	850–1100	

TABLE 3. DIFFUSION IN HOMOGENEOUS ALLOYS—*continued*

Element 1 (purity) at %	Element 2 (purity) at %	A_1*	Q_1*	A_2*	Q_2*	Temp. range (°C)	Ref.
Fe (—)	V (—)		IVb	p.c.			
(γ)	0.53	1.46	68.9	—	—	} 1100–1300	54
(γ)	1.09	0.53	66.2	—	—	(Paramag.)	
(α)	2.11	0.10	61.5				
(—)	(—)		IVa(i)	p.c.			
	18.0	7.0	61.7	3.9	58.5	880–1200	55
Mo (99.4) 2.94	Ni	—	IVb p.c. —	Diffusion of C^{14} 1.0	38	600–900	36
Mo (—) 0 10	U (—)	— —	IIa(ii) — —	p.c. 1.8×10^{-3} 2.5×10^{-3}	27.5 33.0 }	800–1040	57
N V. small	Nb (99.4) ~100	p.c. 0.0086	Va 34.92	—	—	150–300	30
N V. small	Ta ( ) ~100	p.c. Va .0056	37.84	—	—	185–300	30
N V. small	V ( ) ~100	p.c. 0.0092	Va 34.06	—	—	135–280	30
Nb ( ) ~100	O ( ) V. small	p.c. —	Va —	0.0201	26.91	40–150	30
Nb (99.6) 5 10 15	Ti (99.7)	p.c. 1.2×10^{-4} 5.8×10^{-4} 1.5×10^{-3}	IVa(ii) 29.9 36.1 39.3	Diffusion of Fe^{55} 7.9×10^{-3} 11.5×10^{-3} 7.9×10^{-3}	33.1 34.9 35.0	} 850–1300	53
31 54 66 89		9×10^{-3} 8×10^{-3} 0.1 1.0	50.0 64.0 72.0 91.0	— (a) — —	— — —	1050–1800 1200–1800 1500–2000 1700–2200 }	58
Nb (—) 10	U (—)	p.c. —	IIa(ii) —	1.66×10^{-4}	28.2	800–1040	57
Ni (99.95)	W						
	0	1.9	68.0	2.0	71.5	1096–1395 (D_{Ni})	
	1.7 5.3	30.0 58.0	76.5 80.6	2.2 17.0	73.1 80.5	1100–1295 (D_W)	66
	9.2	1.1	70.3	1.4	74.5		

(a) Values read from graphically plotted results.

O V. small	Ta (99.9) 100	p.c. 0.0044	Va 25.45	—	—	40–160	30
O V. small	V (—) 100	p.c. 0.0130	Va 29.01	—	—	70–190	30
Pb (99.99) 100	Tl (99.99)		IVa(i)	p.c. (except 62.6%)			
	0	1.372	26.06	0.511	24.33		
	5.21	1.108	25.75	0.364	23.89		
	10.27	0.880	25.45	0.361	23.83		
	20.2	0.647	25.05	0.353	23.78		
	34.6	0.367	24.53	0.193	23.12		
	50.3	0.231	24.44	0.091	22.52		
	62.4	0.393	25.64	0.101	22.93	} 206–323	59
	62.6 (s.c.)	0.287	25.29	0.126	23.20		
	74.5	0.691	26.83	0.194	23.86		
	76.2	0.862	27.13	0.330	24.48		
	81.8	2.575	28.24	0.957	25.37		
	87.1	17.0	29.71	1.20	25.53		

TABLE 3. DIFFUSION IN HOMOGENEOUS ALLOYS—*continued*

Element 1 (purity) at %	Element 2 (purity) at %	A_1^*	Q_1^*	A_2^*	Q_2^*	Temp. range (°C)	Ref.
Sn	Zr (99·6)						
0		—	IVb	p.c. 5.9×10^{-2}	52	650–827	
1·0		—	—	5.0	62		60
1·85		—	—	2.1×10^{-3}	75	740–827	
2·75		—	—	10	64		
U (—)	Zr (—)		IIa(ii)	p.c.			
10	10	1.26×10^{-4}	22	—	—	800–1040	57
( )	( ) 95		IIa(ii)	p.c.	IVa(i)		
		$D_U^* = 1.5 \times 10^{-9}$	—	$D_{Zr}^* = 3.2 \times 10^{-9}$	—	1000	62
		$D_U^* = 1.85 \times 10^{-9}$	—	$D_{Zr}^* = 4.2 \times 10^{-9}$	—	1050	

REFERENCES TABLE 3

1. R. E. Hoffman, D. Turnbull and E. W. Hart, *Acta metall.*, 1955, **3**, 417.
2. W. C. Mallard, A. B. Gardner, R. F. Bass and L. M. Slifkin, *Phys. Rev.*, 1963, **129**, 617.
3. A. Schoen, *Ph.D. Thesis*, University of Illinois, 1958.
4. N. H. Nachtrieb, J. Petit and J. Wehrenberg, *J. chem. Phys.*, 1957, **26**, 106.
5. R. L. Rowland and N. H. Nachtrieb, *J. phys. Chem.*, 1963, **67**, 2817.
6. E. Sonder, *Phys. Rev.*, 1955, **100**, 1662.
7. R. E. Hoffman, *Acta metall.*, 1958, **6**, 95.
8. D. Lazarus and C. T. Tomizuka, *Phys. Rev.*, 1956, **103**, 1155.
9. R. E. Hoffman and D. Turnbull, *J. appl. Phys.*, 1952, **23**, 1409.
10. C. T. Tomizuka. In the press.
11. T. Heumann and P. Lohman, *Z. Electrochem*, 1955, **59**, 849.
12. W. C. Hagel and J. H. Westbrook, *Trans. metall., Soc. A.I.M.E.*, 1961, **221**, 951.
13. H. A. Domian and H. I. Aaronson, *ibid.*, 1964, **230**, 44.
14. S. D. Gertsricken and I. Y. Dekhtyar, *Proc. 1955 Geneva Conf., 1955*, **15**, 99.
15. ——, *Fizika metall. Metallov.*, 1956, **3**, 242.
16. F. C. Nix and F. E. Jaumot, *Phys. Rev.*, 1951, **83**, 1275.
17. S. D. Gertsricken *et al.*, *Issled. zharpr. Splav.*, 1958, **3**, 68.
18. A. E. Berkowitz, F. E. Jaumot and F. C. Nix, *Phys. Rev.*, 1954, **95**, 1185.
19. J. E. Hilliard, B. L. Averbach and M. Cohen, *Acta metall.*, 1959, **7**, 86.
20. H. B. Huntington, N. C. Miller and V. Nerses, *ibid.*, 1961, **9**, 749.
21. D. Gupta, *Thesis*, University of Illinois, 1961.
22. A. D. Kurtz, B. L. Averbach and M. Cohen, *Acta metall.*, 1955, **3**, 442.
23. J. E. Reynolds, B. L. Averbach and M. Cohen, *ibid.*, 1957, **5**, 29.
24. M. Yanitskaya, A. A. Zhukhavitskii and S. Z. Bokstein, *Dokl. Akad. Nauk SSSR*, 1957, **112**, 720.
25. S. D. Gertsricken and T. K. Yatsenko, *Vop. Fiz.*, 1957, **8**, 101.
26. C. A. Wert and C. Zener, *Phys. Rev.*, 1949, **76**, 1169.
27. R. P. Smith, *Trans. metall. Soc. A.I.M.E.*, 1962, **224**, 105.
28. H. W. Mead and C. E. Birchenall, *J. Metals.*, 1956, **8**, 1336.
29. P. L. Gruzin and E. V. Kuznetsov, *Dokl. Akad. Nauk SSSR*, 1953, **93**, 808.
30. R. W. Powers and M. V. Doyle, *J. appl. Phys.*, 1959, **30**, 514.
31. B. Khomka, *Acta Physica Polonica*, 1963, **24**, 669.
32. ——, *Nukleonika*, 1963, **8**, 185.
33. P. L. Gruzin and B. M. Noskov, " Problems of Metallography and Physics of Metals," **4**, 509 (Moscow, 1955) **and** Aec. tr 2924 p. 355.
34. T. Hirone, N. Kunitomi and M. Sakamoto, *J. phys. Soc. Japan*, 1958, **13**, 840.
35. K. Hirano, R. P. Agarwala, B. L. Averbach and M. Cohen, *J. appl. Phys.*, 1962, **33**, 3049.
36. D. L. Gruzin, Yu. A. Polikarpov and G. B. Federov, *Fizika metall. Metallov.*, 1957, **4**, 94.
37. R. A. Wolfe and H. W. Paxton, *Trans. metall. Soc. A.I.M.E.*, 1964, **230**, 1426.
38. H. W. Paxton and T. Kunitake, *ibid.*, 1960, **218**, 1003.
39. L. I. Ivanov and N. P. Ivanchev, *Izv. Akad. Nauk. SSSR, Otdel. Tekhn. Nauk.*, 1958, **8**, 15.
40. A. Ya Shinyayev, Conference on Uses of Isotopes and Nuclear Radiations, *Met. Metallogr.*, **1958**, p. 299, Moscow.
41. V. Linnenbom, M. Tetenbaum and C. Cheek, *J. appl. Phys.*, 1955, **26**, 932.
42. K. Monma, H. Suto and H. Oikawa, *Nippon Kink. Gakk.*, 1964, **28**, 188.
43. ——, —— ——, *ibid.*, 1964, **28**, 192.
44. A. J. Mortlock and D. H. Tomlin, *Phil. Mag.*, 1959, **4**, 628.
45. J. Hino, C. Tomizuka and C. Wert, *Acta Metall.*, 1957, **5**, 41.
46. A. B. Kuper, D. Lazarus, J. R. Manning and C. T. Tomizuka, *Phys. Rev.*, 1956, **104**, 1536.
47. P. Camagni, *Proc. 2nd Geneva Conf. Atomic Energy*, P/1365, Vol. 20, Geneva, 1958.
48. C. Bassani, P. Camagni and S. Pace, *Il Nuovo Cim.*, 1961, **19**, 393.
49. P. L. Gruzin, B. M. Noskov and V. I. Shirokov, *Dokl. Akad. Nauk SSSR*, 1954, **99**, 247.
50. J. R. MacEwan, J. U. MacEwan and L. Yaffe, *Can. J. Chem.*, 1959, **37**, 1629.
51. Results given are from a ' best line ' through the internal friction data of (−20°–+32°) C. Wert, *J. appl. Phys.*, 1950, **21**, 1196; (9·5°–21·5°) J. D. Fast and M. B. Verrijp, *J. Iron St. Inst.*, 1954, **176**, 24; (18·5°–59°) W. R. Thomas and G. M. Leak, *Phil. Mag.*, 1954, **45**, 656; (91°–161°) L. Guillet and P. Gence, *J. Iron St. Inst.*, 1957, **186**, 223.
52. M. E. de Morton, *J. appl. Phys.*, 33, 2768, 1962, **33**, 2768.
53. R. F. Peart and D. H. Tomlin, *Acta metall.*, 1962, **10**, 123.
54. M. S. Zelinski, B. M. Moskov, P. V. Pavlov and E. V. Shitov, *Physics Metals Metallogr.*, 1959, **8** (5), 79 and *Fiz. Metall. Metallov.*, 1959, **8**, 725.
55. J. Stanley and C. Wert, *J. appl. Phys.*, 1961, **32**, 267.
56. I. A. Naskidashvili, *Soobshoheniya Akad. Nauk. Gauzin. SSSR*, 1955, **16**, 509.
57. Y. Adda and A. Kirianenko, *J. Nucl. Mater.*, 1962, **6**, 135.
58. G. B. Gibbs, D. Graham and D. H. Tomlin, *Phil. Mag.*, 1963, **8**, 1269.
59. H. A. Resing and N. H. Nachtrieb, *Physics Chem. Solids*, 1961, **21**, 40.
60. V. S. Lyashenko, V. N. Bykov and L. V. Pavlinov, *Physics Metals Metallogr.*, 1959, (3) **8**, 40.
61. T. Heumann and F. Heinemann, *Z. Electrochem.*, 1956, **60**, 1160.

62. Y. Adda, C. Mairy and J. M. Andreu, *Revue Métall. Paris*, 1960, **57**, 549.
63. A. B. Gardner, *Thesis*, University of N. Carolina, 1964.
64. R. J. Borg, *J. appl. Phys.*, 1963, **34**, 1562.
65. S. D. Gertsricken, and I. Ya. Dekhtyar, *Proc. 1955 Geneva Conf. Peaceful Uses of Atomic Energy*, 1955, **15**, 124.
66. K. Monma, H. Suto and H. Oikawa, *Nippon Kink. Gakk.*, 1964, **28**, 197.
67. D. A. Leak and G. M. Leak, *J. Iron St. Inst.*, 1958, **189**, 256.
68. J. Askill, *Thesis*, University of Reading, 1964.

TABLE 4. CHEMICAL DIFFUSION COEFFICIENT MEASUREMENTS

Element 1 at %	Element 2 at %	A	Q	D	Temp. range (°C)	Method	Ref.
A V. small	Ag $\sim$100	0·12	33·6		600–800	IIIb(ii)	1
A V. small	Mg $\sim$100	10^4	52		330–540	IIIb(ii)	104
Ag 0·5 1·0 1·5 2·0 2·5 3·0 3·5 6·5 8·5	Al	0·21 0·30 0·33 0·55 0·78 1·50 3·0 11·0 16·0	28·8 29·5 29·8 30·8 31·4 32·5 33·7 37·0 38·0	A few partial D_{Ag}'s and D_{Al}'s calculated. Very roughly $\tilde{D} = D_{Ag} \sim 2D_{Al}$	500–595	IIa(i)	2
Ag 0–8·77 45·0 54·0 64·0 75·0 85·0 50·8	Au	0·0242 0·14	37·0 41·7	$4\cdot7 \times 10^{-9}$ $4\cdot1 \times 10^{-9}$ $3\cdot7 \times 10^{-9}$ $2\cdot8 \times 10^{-9}$ $1\cdot9 \times 10^{-9}$	806–1017 940 763–965	IIa(ii) IIIa(i) IIa(i)	3 4 5
Ag	Cd 0–5 0–25	$4\cdot7 \times 10^{-3}$	31·3	Figure 3	650–895 627, 727 and 780	IIa(ii) and IIIb(ii) IIa(i)	6 51
Ag 0–3	Cu	0·012	35·6		717–867	IIa(ii)	6
Ag	**H** o–sol. limit	$2\cdot82 \times 10^{-3}$ (D indep. of conc.)	7·5		338–600	IIIb(ii)	7
Ag $\sim$100	**Kr** V. small	1·05	35·0		500–800	IIIb(ii)	8
Ag $\sim$100	**Ne** V. small	2·28	59·4			IIIb(ii)	105
Ag	**O** o–sol. limit	$3\cdot66 \times 10^{-3}$	11·0		412–862	IIIb(ii)	9
Ag 0–0·12	Pb	$7\cdot4 \times 10^{-2}$	15·2		220–285	IIa(ii)	10
Ag $\sim$100	**Xe** V. small	0·036	37·5		500–800	IIIb(ii)	13
Ag (β) 50 40–55 (α) 26·8 27·6	**Zn**	0·0164	16·5	$D_{Zn} \sim$ 3 to 4D_{Ag} Figure 4 $8\cdot7 \times 10^{-9}$ $2\cdot45 \times 10^{-9}$ $D_{Zn} \sim 1\cdot5$–2·2D_{Ag}	400–610 at 700 at 650	{ IIc { IIa(i) IIIa(i)	11 12
Al	Be 0·015 0·022 0·03	52 126 550	39·0 40·3 43·1		500–635	IIa(iii) and IIc	14
Al 0–12·5 $\sim$25	Cu 0–0·215 (β)	0·29 0·19	31·12 27·5	Figure 5	505–635 890–995 646–750	IIa(iii) IIa(i) IIc	15 16 17
Al o–18 o–$\sim$52	Fe (α) (α)	30·1	56·0	Figure 6 $D_{Al} = 8\cdot6 \times 10^{-10}$ $D_{Fe} = 4\cdot4 \times 10^{-10}$	850 950–1100	IIIa(i) IIa(ii)	18 120

TABLE 4. CHEMICAL DIFFUSION COEFFICIENT MEASUREMENTS—*continued*

Element 1 at %	Element 2 at %	A	Q	D	Temp. range (°C)	Method	Ref.
Al	**H** o-sol. limit	0·21	10·9		470–590	IIIb(ii)	19
Al ~100	**He** V. small	3·0	36·5			IIIb(ii)	105
Al	Li o-sol. limit	4·5	33·3		417–597	IIa(ii)	20
Al	Mg 0·27–2·2	Results scattered. Most fall in the band of *Figure 7* for this conc. range.			400–550		eg 21
Al	Mn 0·02–0·15			*Figure 8*	600–650	IIa(iii)	21
Al	Na 0–0·002	1·1	32·0		550–650	IIIb(i)	22
Al 0–0·7	Ni	1·87	64·0		1100–1280	IIa(ii)	23
Al	Si 0–0·5 / 0–0·7	0·90	30·5	*Figure 9*	465–600 / 450–580	IIa(iii) IIa(iii)	24 / 21 and 25
Al 2·0 / 12·0 / 10·0 / 3·8	Ti (β) / (α) / (β)	$1·4 \times 10^{-5}$ / $9·0 \times 10^{-5}$ / $1·6 \times 10^{-5}$	21·9 / 25·5 / 23·7	$\tilde{D}$ increases linearly with c. $D_{Al} = 14·11 \times 10^{-9}$ $D_{Ti} = 4·61 \times 10^{-9}$	983–1250 / 834–900 / 1250	IIa(i)	26
Al	Zn	$\tilde{D}$ @ 330°	360°	400° 440°	485°	540°	27
	~0	1·84	3·98	12·7 49·2	149	610	
	9·0	1·95	4·85	19·3 69·6	174	610	
	18·1	3·64	6·12	20·0 74·8	2·2	—	
	37·6	—	1·10	— 51·1	—	—	
		All D's $\times 10^{-11}$					
	To very good approximation			$D_{Zn} = \tilde{D}/c_{Al}$ (c_{Al} = fractional at. conc.)	330–540	IIa(i)	
Au ~0–1·5	Cu 0–~10	$2·10^{-4}$ $1·2 \times 10^{-3}$	29·8 28·5		650–970 540–740	IIb IIb	6 6
0·5				$0·39 \times 10^{-10}$			
1·0				$0·45 \times 10^{-10}$			
2·0				$0·54 \times 10^{-10}$			
5·0				$0·75 \times 10^{-10}$			
10·0				$1·0 \times 10^{-10}$			
15·0				$1·3 \times 10^{-10}$			
20·0				$3·0 \times 10^{-10}$	750	IIa(i)	52
30·0				$6·0 \times 10^{-10}$			
40·0				$10·0 \times 10^{-10}$			
60·0				$13·0 \times 10^{-10}$			
80·0				$15·0 \times 10^{-10}$			
90·0				$16·0 \times 10^{-10}$			
Au	Fe 0–18·3	$1·16 \times 10^{-4}$	24·4		750–1000	IIa(ii)	28
Au	**H** o-sol. limit	$5·6 \times 10^{-4}$	5·64		500–940	IIIb(ii)	29
Au	In			$\tilde{D}$ at 142° $\tilde{D}$ at 151° ($\times 10^{-12}$)			30
3				0·47 24·0			
33	(Au In_3)			7·0 29·0			
50	(Au In)			2·6 6·6		—	
69	(Au_9 In_4)			5·8 9·8		IIa(i)	
80	(Au_4 In)			0·49 0·68		(Values of D_{In} listed, calculated assuming $D_{Au} \ll D_{In}$)	
91				0·24 0·28			
Au	Ni						
2		$4·3 \times 10^{-2}$	41·4				
10		$3·9 \times 10^{-2}$	41·5				
20		$9·5 \times 10^{-2}$	43·7				
30		$7·8 \times 10^{-2}$	43·9				
40		$2·2 \times 10^{-2}$	42·0				
50		$1·3 \times 10^{-2}$	42·4		850–975	IIa(i)	31
55		$5·9 \times 10^{-3}$	41·7				
60		$6·8 \times 10^{-2}$	48·8				
65		$1·4 \times 10^{3}$	73·0				
70		$2·0 \times 10^{7}$	96·2				
75		$6·2 \times 10^{8}$	105·0				
98		$1·8 \times 10^{4}$	85·2				

TABLE 4. CHEMICAL DIFFUSION COEFFICIENT MEASUREMENTS—*continued*

Element 1 at %	Element 2 at %	A	Q	D	Temp. range (°C)	Method	Ref.
Au 0–~0.09	Pb 0	0.35	14.0		113–300	IIa(ii)	32
Au	Pd 0–17.1	1.13×10^{-3}	37.4		727–970	IIb	33
Au 98 96 94 92 90 88 86 84 82 80 2–8 95	Pt	0.62 1.0 0.73 0.67 0.60 0.58 0.53 0.52 0.47 0.43 0.37 $A_{Au} = 0.32$ $A_{Pt} = 0.09$	54.6 56.0 55.4 55.4 55.2 55.2 55.1 55.2 55.1 55.0 62.6 $Q_{Au} = 45.6$ $Q_{Pt} = 54$		925–1055	IIa(i)	34
B 0–0.0095 0–0.02	Fe (α) (γ)	10^6 0.002	62.0 21.0		700–835 950–1300	IIIa(ii) IIIa(ii)	35 36
		(Analysis of 'Fe' for both experiments (in wt%) 0.0038B, 0.43C, 1.64Mn, 0.02P, 0.019S, 0.37Si, 0.04Cr, 0.01Ni, 0.01Mn)					
Ba o–sol. limit	U (γ)	0.112	40.8		850–1040	IIc	85
Be 0–~15 ~33 ~48 ~75 ~33	Cu (α) (β) (γ(β')) (δ) (β)	0.19 0.084 0.054 0.0012 $A_{Be} = 0.035$ $A_{Cu} = 0.045$	41.5 27.5 31.0 33.0 $Q_{Be} = 29$ $Q_{Cu} = 25$		550–884 650–884 550–884 550–884	IIa(i)	37
Be	Fe 0–0.2	1.0 Very little variation of $\tilde{D}$ with c	54.0		800–1100	IIa(i)	38
Be	H o–sol. limit (<0.0075)			$3 \cdot 10^{-9}$	850–900	IIc	86
Bi 0–2.0	Pb	0.018	18.4		220–285	IIa(ii)	39
C 0.49	Co	1.765	41.5		850–1100	IIIb(ii)	41
C 0.48 0.48 0–0.7 (wt%)	Co 78.5 89.4 0 5.8 10.6 20.2 (γ)	0.472 0.442 $0.04 + 0.08c$ $0.04 + 0.08c$ $0.03 + 0.1c$ $0.03 + 0.06c$ $(c = wt\% C)$	37.5 37.5 31.35 30.5 29.9 28.85	Fe	850–1100 850–1112 1000–1200	IIIb(ii) IIa(i)	41 146
C 0–0.1 0.46 1.0 2.0 3.0 4.0 5.0 6.0	Fe (α) (γ) (γ)	$\tilde{D}_c = 0.008 \exp(-19{,}800/RT) + 2.2 \exp(-29{,}300/RT)$ $\tilde{D}_c$ indep. of conc. 0.668 0.36 0.27 0.20 0.14 0.10 0.07	37.46 36.0 34.5 33.3 31.9 30.1 28.5		721–845 925–1100 750–1300	IIa(ii) IIIb(ii) IIa(i)	40 41 42
C 0.4 (wt %)	Fe (γ)	Mn 0 1.0 12.1 19.2 → 0.07 0.08 0.19 0.41	31.35 31.60 33.9 36.1		1000–1200	IIa(i)	148
C 0–0.7 (wt %)	Fe (γ)	Ni 0 3.9 9.2 17.3 → $0.04 + 0.08c$ $0.03 + 0.1c$ $0.03 + 0.1c$ $0.02 + 0.1c$ $(c = wt\% C)$	31.35 31.0 30.55 29.8		1000–1200	IIa(i)	147

TABLE 4. CHEMICAL DIFFUSION COEFFICIENT MEASUREMENTS—*continued*

Element 1 at %	Element 2 at %	A	Q	D	Temp. range (°C)	Method	Ref.
C Sol. soln. range	Fe (γ) Si 0–2·35			Figure 22	880–950	IIc	145
C 0–sol. limit	Ni	2·48	40·2		730–1020	IIIb(ii)	43
C ~7·2	Th		~38	$\left\{\begin{array}{l}7\cdot4\times10^{-9}\\2\cdot8\times10^{-8}\\5\cdot7\times10^{-8}\end{array}\right.$	$\left.\begin{array}{l}1000\\1100\\1200\end{array}\right\}$	IIIa(i)	44
C 0·14–sol. limit 0·14–sol. limit	Ti (α) (β)	5·06 108·0	43·5 48·4	$\tilde{D}$ ind. of conc. $\tilde{D}$ ind. of conc.	736–835 950–1150	IIc IIc	$\left.\begin{array}{l}\\\end{array}\right\}45$
Cd 0–0·5	Cu	$3\cdot5\times10^{-3}$	29·2		500–850	IIa(ii)	6
Cd 0–1·0	Pb	$1\cdot85\times10^{-3}$	15·4		167–252	II	46
Cd	**Hg** 0–4·0	2·57	19·6		156–202	III(ii)	46
Ce 0–sol. limit	U (γ)	3·92	66·4		800–1000	IIc	85
Co	Cr 0–15·2 0–40	0·084 0·443	60·6 63·6		1000–1300 1000–1370	IIa(ii) IIa(iii)	47 48
		(*D* reported f(c), but concentration dependence very slight)					
Co	Mo 0–10	0·231	62·8		1000–1300	IIa(ii)	47
Co	Ni 5–90			Figure 30	1150–1420	IIa(i)	144
Co	V 0–17	0·021	53·0		1100–1300	IIa(ii)	47
Co	W 0–5	0·008	56·9		1100–1300	IIa(ii)	47
Cr 10–20	Fe (α)	1·48	54·9	$\tilde{D}$ indep. of c (In the range 10–1%, $\tilde{D}$ increases by ~30%.) $D_{Cr}\sim1\cdot5\,D_{Fe}$	823–1440	IIa(i)	49
37 42 52 62 72 81 91	(α)			$\left\{\begin{array}{l}80\times10^{-10}\\56\times10^{-10}\\27\cdot8\times10^{-10}\\11\cdot7\times10^{-10}\\7\cdot1\times10^{-10}\\4\cdot9\times10^{-10}\\4\cdot64\times10^{-10}\end{array}\right.$	1250	IIa(i)	50
0–7·1	(γ)	0·0012	52·2		900–1200	IIa(ii)	47
Cr (γ) Cr$_2$Nb (α)	Nb 2 34 38 (a) 90 95	$\left\{\begin{array}{l}34\cdot0\\26\cdot0\\24\cdot0\\2\cdot7\\0\cdot31\end{array}\right.$	97·9 94·0 94·8 (b) 92·0 86·0 $\left.\begin{array}{l}\\\\\end{array}\right\}$		1100–1624 1251–1624 1100–1624	$\left.\begin{array}{l}\\\\\end{array}\right\}$IIa(i)	140
Cr 0–12	Ni	0·604	61·44		1000–1300	IIa(ii)	47
Cr 9	Ti (β)			$\left.\begin{array}{l}3\cdot6\times10^{-9}\\D_{Ti}=2\cdot8\times10^{-9}\\D_{Cr}=3\cdot7\times10^{-9}\end{array}\right\}$	985	IIa(i)	53
Cr 0–sol. limit	U (γ)	0·7	34		900–1000	IIc	54

(a) The compound Cr$_2$Nb is reported to have a 5% solubility range.
(b) Arrhenius plots not always linear. $\tilde{Q}$ and $\tilde{A}$ derived from measurements at the higher temperatures.

TABLE 4. CHEMICAL DIFFUSION COEFFICIENT MEASUREMENTS—*continued*

Element 1 at %	Element 2 at %	A	Q	D	Temp. range (°C)	Method	Ref.
Cu	H o-sol. limit	0·011	9·2	D ind. of conc.	430–650	IIIb(ii)	19
Cu 0–95	Ni		Figure 11	Figures 10 and 12	947–1054	IIa(i)	16, 55 and 57
89·9				$D_{Cu}/D_{Ni} = 1·3$	922–1050	IIa(i)	55
5·0				$1·7 \times 10^{-13}$			
10·0				$1·7 \times 10^{-13}$			
20·0				$2·0 \times 10^{-13}$			
40·0				$3·0 \times 10^{-13}$			
60·0				$5·0 \times 10^{-13}$			
70·0				$9·0 \times 10^{-13}$	750	IIa(i)	56
80·0				$16·0 \times 10^{-13}$			
90·0				$33·0 \times 10^{-13}$			
95·0				$43·0 \times 10^{-13}$			
98·0				$51·0 \times 10^{-13}$			
99·0				$52·0 \times 10^{-13}$			
99·5				$70·0 \times 10^{-13}$			
Cu	O o-sol. limit in eq. with Cu_2O	0·0748	46		600–950	IIIb(ii)	58
Cu 0–100	Pd			Figure 13	878–1038	IIa(i)	55
Cu 0–13·9	Pt	0·049	55·7		1040–1400	IIa(ii)	28
Cu	Si 0	0·037	40·0				
	4	0·41	48·2		700–800	IIa(i)	59
	8	18·6	53·8				
Cu (α)	Sn 1–5			Figure 14	733–812	IIa(i)	16
				$7·7 \times 10^{-10}$	428		
				$3·6 \times 10^{-9}$	441		
(δ)	20·5	{ D_{Cu} D_{Sn} }	Figure 15	$1·0 \times 10^{-8}$	458	IIc	
				$1·1 \times 10^{-7}$	507		60
				$1·65 \times 10^{-7}$	545		
				$1·3 \times 10^{-7}$	572		
(γ)	15–22	{ D_{Cu} D_{Sn} }	Figure 17	Figure 16	706·5	IIa(i)	
(ε)	24·5			$4·3 \times 10^{-12}$	180	IIc	
				$3·7 \times 10^{-11}$	210		
(η)	45·0			$2·7 \times 10^{-12}$	180	IIc	
				$2·2 \times 10^{-11}$	210		
Cu	Zn 1	0·056	40·0				
	5	0·062	40·0		780–915		
	10	0·083	39·5				
(α)	16	0·095	38·0			IIa(i)	61
	20	0·09	36·5				
	25	0·031	32·5		724–915		
	28	0·016	29·7				
		D_{Cu} and D_{Zn}—Figures 18, 19 and 20					
	10	0·13	40·8				
	15	0·21	40·8				
(α)	20	0·36	40·8		700–910	IIIa(i)	62
	28	1·7	41·3				
	28 { D_{Zn} D_{Cu}	2·1	41·2				
		0·81	42·7				
(β)	46			$D_{Zn}/D_{Cu} = 2·4$ to $3·6$	600–800		
(β)	44 to 48	0·018 to 0·013	19·9 to 18·2		500–800	IIa(i)	63
	59	$2·45 \times 10^{-2}$	23·4				
	60	$2·44 \times 10^{-2}$	22·9				
	61	$1·71 \times 10^{-2}$	21·9				
	62	$2·45 \times 10^{-2}$	21·9		375–650		
(γ)	63	$1·14 \times 10^{-2}$	20·1				
	64	$0·99 \times 10^{-2}$	19·3				
	65	$0·62 \times 10^{-2}$	17·7		425–650	IIIa(i)	64
	65·5	$0·19 \times 10^{-2}$	15·5		525–650		
	66·5	$0·28 \times 10^{-2}$	15·3				
				D_{Zn}/D_{Cu}			
	65–66			9·4	375–475		
(γ)	67			11·4	525		
	68			8·6	575		
	68			5·7	650		

TABLE 4.   CHEMICAL DIFFUSION COEFFICIENT MEASUREMENTS—*continued*

Element 1 at %	Element 2 at %		A	Q	D	Temp. range (°C)	Method	Ref.
Fe (α)	**H** 0–0·08		$1·4 \times 10^{-3}$	3·2	Indep. of c.	200–780	IIIa(ii)	7
(α)	0–sol. limit		$0·93 \times 10^{-3}$	2·7		200–774	IIIa(ii)	65
			At $T < \sim 200°C$ $\tilde{D}$ is usually less than expected from extrapolation of higher T results and apparently depends on sample history. See Ref. 65					
(γ)			0·011	9·95			Ib	66
Fe	Mn	C (wt-%)						
(γ)	4	0·02	0·57	66·2 ⎫		1050–1450 ⎫	IIa(i) and (ii)	67
	14	0·02	0·54	65·4 ⎬		⎬		
	4	1·25	0·51	61·2 ⎬		1000–1250 ⎭		
	14	1·25	0·52	61·0 ⎭				
			Empirically $\tilde{D}$ may be represented ($\pm 20\%$) over the ranges 0–20% Mn, 0–1·5 wt% C by $\tilde{D} = (0·486 + 0·011$ wt% Mn$)(1 + 2·53$ wt% C$) \exp(-66{,}000/RT)$					
(γ)	0–60				*Figure 21*	1200	IIa(i)	67
Fe (γ)	Mo 0–0·59		0·068	59·0		1150–1260	IIa(ii) ⎫	89
			(Addition of 0·4 wt% C increases A to 0·091)				⎬	
(α)	1·9–3·6		3·467	57·7		930–1260	IIa(ii) ⎭	
Fe (α + δ)	**N** 0–sol. in eq. with 0·95 atm. N₂		$7·8 \times 10^{-3}$	18·9		500–850 and 1410–1470	IIIa(i) and IIIb(ii)	68 69
(γ)	0–sol. in eq. with 0·95 atm. N₂		0·91	40·26		950–1350	IIIa(i) and IIIb(ii)	70 and 71 69
Fe (α)	Ni 1				$\begin{cases} 2·9 \times 10^{-14} \\ 2·75 \times 10^{-12} \end{cases}$	700 800		
	10		5·3	76·2				
	20		8·9	76·0				
	30		15·0	75·9				
	40		24·5	75·8				
(γ)	50		41·5	75·7		1000–1290	IIa(i)	141
	60		58·5	75·7				
	70		38·5	73·4	$D_{Fe} > D_{Ni}$ for $c < \sim 60\%$ Ni			
	80		44·5	73·8	$D_{Fe} < D_{Ni}$ for $c > \sim 60\%$ Ni			
	90		49·5	74·8				
			Effect of a pressure of 40 kb is to decrease $\tilde{D}_\gamma$ by one order of magnitude					
Fe	Ni	C (wt-%)						
(γ)	4	0·03	0·44	67·7 ⎫		1100–1450 ⎫	IIa(i) and (ii)	72
	16	0·03	0·51	67·3 ⎬		⎬		
	4	0·6	0·46	65·5 ⎬		1050–1300 ⎭		
	16	0·6	0·42	64·5 ⎭				
			Empirically, $\tilde{D}$ may be represented ($\pm 20\%$) over the ranges 0–20% Ni, 0–1·5 wt% C by $\tilde{D} = (0·344 + 0·012$ wt% Ni$) \times (1 + 2·3$ wt% C$) \exp(-67{,}500/RT)$					
Fe (α and δ)	P Sol. soln. range		$\sim 2·9$	~ 55		850–875 and 1410–1458	IIa(i)	79
(γ)	Sol. soln. range		28·3	69·8		1250–1350		
Fe (α)	S Sol. soln. range		1·68	48·9		750–900	IIIa(i)	73
(α and δ)	Sol. soln. range		1·35	48·4		750–900 and 1400–1450	IIa(i)	73 and 74
(γ)	Sol. soln. range		2·42	53·4		1200–1352	IIa(i)	74
Fe (α)	S Sol. soln. range	Si 6·2	2·68	49·7		900–1300	IIIa(i)	73
Fe (α)	Si 4·5–7·1		0·44	48·0		1095–1350	IIa(ii)	75
(α)	0–~6				$1·1 \times 10^{-7}$	1435	IIa	76
(α)	6–14				*Figure 23*	1150	IIc	80
(γ)	0–2				$\begin{cases} 4 \times 10^{-10} \\ 1·7 \times 10^{-9} \end{cases}$	1206 1293 ⎬	IIa(ii)	75

TABLE 4.　CHEMICAL DIFFUSION COEFFICIENT MEASUREMENTS—*continued*

Element 1 at %	Element 2 at %	A	Q	D	Temp. range (°C)	Method	Ref.
Fe	Sn			$\begin{cases} 9 \cdot 7 \times 10^{-10} \\ 2 \times 10^{-9} \\ 3 \cdot 9 \times 10^{-9} \\ 7 \cdot 6 \times 10^{-9} \end{cases}$	$\left.\begin{array}{l} 950 \\ 1000 \\ 1050 \\ 1100 \end{array}\right\}$	II	81
	0–100	—	~46				
Fe (α) (γ)	Ti ~0·7–~0·3 0–~0·7	3·15 0·15	59·2 60·0		1075–1225 1075–1225	IIc IIa(i) $\Big\}$	77
Fe o-sol. limit	U (γ)	1·3	32·0		790–1000	IIc	54
Fe (α) 1 atm.	V $\begin{cases} 0 \cdot 7 \\ 5 \cdot 0 \\ 10 \cdot 0 \\ 15 \cdot 0 \\ 20 \cdot 0 \\ 25 \cdot 0 \\ 30 \cdot 0 \end{cases}$	$\begin{array}{l} 0 \cdot 61 \\ 3 \cdot 9 \\ 1 \cdot 1 \\ 0 \cdot 70 \\ 0 \cdot 71 \\ 0 \cdot 63 \\ 0 \cdot 59 \end{array}$	$\left.\begin{array}{l} 63 \cdot 8 \\ 56 \cdot 9 \\ 53 \cdot 5 \\ 52 \cdot 5 \\ 52 \cdot 8 \\ 52 \cdot 8 \\ 53 \cdot 1 \end{array}\right\}$		$\left.\begin{array}{l} 950\text{–}1250 \\ \end{array}\right\}$		
	10·0 (20 and 40 kb pressure)			*Figure 24*			
(γ)	0·7 (1, 20 and 40 kb)			*Figure 25*	950–1300	IIa(i)	78
(γ)	2·0 (40 kb)			$\begin{cases} 5 \cdot 8 \times 10^{-12} \\ 2 \cdot 5 \times 10^{-11} \\ 2 \cdot 9 \times 10^{-11} \\ 1 \cdot 1 \times 10^{-10} \\ 2 \cdot 6 \times 10^{-10} \end{cases}$	$\begin{array}{l} 1100 \\ 1162 \\ 1201 \\ 1275 \\ 1350 \end{array}$		
(γ)	3·0 (40 kb)			$\begin{cases} 6 \cdot 3 \times 10^{-12} \\ 2 \cdot 0 \times 10^{-11} \\ 3 \cdot 2 \times 10^{-11} \\ 1 \cdot 6 \times 10^{-10} \\ 1 \cdot 4 \times 10^{-10} \\ 2 \cdot 8 \times 10^{-10} \end{cases}$	$\begin{array}{l} 1100 \\ 1162 \\ 1201 \\ 1275 \\ 1292 \\ 1350 \end{array}$		
Fe 0–0·13	W	11·5	142		1927–2527	IIIb(ii)	82
	0–1·3 0–1·2 0–3·4			$\begin{array}{l} 3 \cdot 7 \times 10^{-10} \\ 2 \cdot 4 \times 10^{-9} \\ 1 \cdot 0 \times 10^{-9} \end{array}$	$\left.\begin{array}{l} 1280 \\ 1330 \\ 1330 \end{array}\right\}$	IIa	83
Ge	**H** Range of c in eq. with H_2 gas over p. range 10–76 cm Hg	$2 \cdot 72 \times 10^{-2}$	8·7		800–910	Ic	84
Ge ~100	**He** V. small	$6 \cdot 1 \times 10^{-2}$	16·0		795–872	Ic	106
H 5–17	Nb	0·0215	9·37		600–700	IIIb(i)	87
H 0–~0·25 o-sol. in eq. with H_2 at 32 cm Hg and 870°C	Ni	0·0045 0·0107	8·6 10·1 $\Big\}$	D apparently indep. of conc.	$\begin{cases} 380\text{–}990 \\ 160\text{–}500 \end{cases}$	IIIb(ii) IIIb(ii)	88 90
H 0–~2	Pd	$1 \cdot 3 \times 10^{-4}$ $4 \cdot 3 \times 10^{-3}$	5·0 5·62		231–334(?) 200–700	Ic Ib	91 92 and 93
		Permeation of H through Pd very much affected by sample history, contamination, etc. See 91–95					
H 0–~10^{-2}	Si	$9 \cdot 4 \times 10^{-3}$	11·0		1090–1200	Ic	106
H Sol. soln. range	Th	$2 \cdot 92 \times 10^{-3}$	9·75		300–900	IIIa(i) and IIIb(ii)	96
H Sol. soln. range Sol. soln. range	Ti (α) (β)	$1 \cdot 8 \times 10^{-2}$ $1 \cdot 95 \times 10^{-3}$	12·38 6·64		500–824 600–1000	IIIa(i) IIIb(i)	97 97
H (n.s.) (n.s.)	U (α) (β) (β) (γ)	0·0195 $3 \cdot 3 \times 10^{-4}$ $1 \cdot 5 \times 10^{-3}$	11·1 3·6 11·4	$\begin{cases} 7 \cdot 3 \times 10^{-5} \\ 1 \cdot 0 \times 10^{-4} \end{cases}$	$\begin{array}{l} 390\text{–}630 \\ 700 \\ 724 \\ 698\text{–}750 \\ 800\text{–}970 \end{array}$	IIIb(ii) IIIb(ii) IIIb(ii) IIIb(ii) IIIb(ii)	98 99

TABLE 4. CHEMICAL DIFFUSION COEFFICIENT MEASUREMENTS—*continued*

Element 1 at %	Element 2 at %	A	Q	D	Temp. range (°C)	Method	Ref.
H Sol. soln. range Sol. soln. range 0–41	Zr (α) (α) (α 'Zircalloy 2') (β) (β)	$4 \cdot 15 \times 10^{-3}$ 7×10^{-4} $2 \cdot 17 \times 10^{-3}$ $5 \cdot 32 \times 10^{-3}$ $7 \cdot 37 \times 10^{-3}$	9·47 7·06 8·38 8·32 8·54	Ind. of c	450–700 300–600 260–560 760–1010 870–1100	IIa(i) IIIa(i) IIIa(i) III(i) IIa(i)	100 101 102 103 100
He $0 - \sim 6 \cdot 10^{-9}$	Mg	60·0	36·0		400–575	IIIb(ii)	104
He $0 - \sim 4 \cdot 10^{-10}$	Si	0·11	29		1170–1207	Ic	106
He 0–0·13	Th (α)	$10^{-3} - 10^{-4}$	38		900–1450	IIIb(ii)	107
He 16	Ti (α)	$1 \cdot 1 \times 10^{-9}$	16·1		615–720	IIIb(ii)	117
Hf	**O** Sol. soln. range, in eq. with oxide	0·66	50·8		500–1050	IIc and IIIb(i)	108
Hg 0–4	Pb	0·35	19		177–197	IIa(ii)	46
In 0–1 0–3	Pb			$\left\{ \begin{array}{l} 2 \cdot 3 \times 10^{-10} \\ 3 \cdot 5 \times 10^{-11} \\ 3 \cdot 5 \times 10^{-10} \end{array} \right.$	285 252 320 $\Big\}$	IIa(ii)	46
Ir 3 24 50 60 90	W (α) (σ) (ε) (ε) (β)	$3 \cdot 9 \times 10^{4}$ $2 \cdot 4 \times 10^{-6}$ 15·0 15·0 $1 \cdot 1 \times 10^{3}$	$\left. \begin{array}{l} 170 \\ 60 \cdot 8 \\ 120 \cdot 4 \\ 120 \cdot 4 \\ 145 \cdot 4 \end{array} \right\}$	(a)	1300–2110	IIa(i)	140
La Sol. soln. range	U (γ)	117·0	55·7		850–1090	IIc	109
Li V. small	Si	$2 \cdot 5 \times 10^{-3}$	15·1		800 and 1350	IIIb(ii)	110
Li n.s., but probably small	W	5·0	41·5		1090–1230	IIIb(ii)	111
Li Sol. soln. range	Zr	0·73	33·7		775–850	IIa(ii)	143

(a) Arrhenius plots non-linear. $\bar{Q}$ and $\bar{A}$ calculated from measurements at the highest temperatures.

Element 1 at %	Element 2 at %	A	Q	D	Temp. range (°C)	Method	Ref.
Mg $0-<1$	Ni	0·44	56		1050–1300	IIa(ii)	112
Mg 0·26 1·0 2·0 3·0 4·1 0·26 1·9	Pb			$2 \cdot 5 - 3 \cdot 7 \times 10^{-10}$ $6 \cdot 9 \times 10^{-10}$ $8 \cdot 6 \times 10^{-10}$ $1 \cdot 1 \times 10^{-9}$ $6 \cdot 4 - 7 \cdot 8 \times 10^{-10}$ $9 \cdot 4 \times 10^{-10}$ $\left\{ \begin{array}{l} 1 \cdot 2 \times 10^{-10} \\ 1 \cdot 3 \times 10^{-10} \end{array} \right.$	250 $\Big\}$ 250 250 270 220 270 $\Big\}$	IIa(iii) IIa(i) IIa(iii) IIa(iii) IIa(iii)	113
Mg 0·01 0·56 1·12 1·7	Pu	in units 10^{-11}	$\bar{D}$ at $T = 420°$ 6·1 3·5 3·1 2·1	$475°$ 25·0 11·3 9·3 13·0	$534°$ 130 46·3 49·7 23·5 $\Big\}$	IIa(i)	118
Mg 0·025	U			$\left\{ \begin{array}{l} 1 \cdot 2 \times 10^{-11} \\ 3 \cdot 3 \times 10^{-11} \end{array} \right.$	400 500 $\Big\}$	IIa(i)	118

TABLE 4.   CHEMICAL DIFFUSION COEFFICIENT MEASUREMENTS—*continued*

Element 1 at %	Element 2 at %	A	Q	D	Temp. range (°C)	Method	Ref.
Mn 0–4	Ni	7·5	67·1		1100–1300	IIa(ii)	23
Mn 8	Ti (β)	$1 \cdot 10^{-3}$	35·2	(Very small dep. of $\tilde{D}$ on c in range 2–13%)	830–1190	IIa(i) and IIIa(ii)	26
Mo ~0 50 ~100	Nb	$1 \cdot 10^{3}$ $1 \cdot 10^{3}$ $1 \cdot 10^{3}$	132 137 138	$D_{Nb} \sim$ 3 to 6 × D_{Mo}	1800–2165	IIa(i)	115
Mo 0–0·93 0–9	Ni	3·0 0·853	68·9 64·4		1150–1400 1000–1300	IIa(ii) IIa(ii)	112 47
Mo 1·0	Ti (β)	$1 \cdot 10^{-5}$ (Very small dep. of D on c in range 1–6%)	24		938–1248	IIa(i)	26
4·0	(β)			$D_{Mo} = 3 \cdot 95 \times 10^{-9}$ $D_{Ti} = 6 \cdot 72 \times 10^{-9}$	1250		26
0–10	(β)	$1 \cdot 3 \times 10^{-4}$	33·1		900–1300	IIa(ii)	114
Sol. soln. range	(α)	$3 \cdot 5 \times 10^{-8}$	28·4		600–800	IIa(ii)	114
Mo 2 4 6 8 10 12 16 20 24 26	U (γ)	2·2 0·58 20·0 16·0 28·0 3·2 0·096 $3 \cdot 10^{-3}$ $4 \cdot 5 \times 10^{-4}$ $2 \cdot 1 \times 10^{-4}$	47·5 45·8 53·0 55·0 56·8 52·2 45·7 39·4 38·5 34·0		850–1050	IIa(i)	116
6·0 8·0 10·0				D_U D_{Mo} $3 \cdot 4 \times 10^{-9}$ $5 \cdot 2 \times 10^{-10}$ $1 \cdot 4 \times 10^{-8}$ $2 \cdot 1 \times 10^{-9}$ $1 \cdot 6 \times 10^{-8}$ $5 \cdot 0 \times 10^{-9}$ $3 \cdot 4 \times 10^{-8}$ $1 \cdot 3 \times 10^{-8}$	850 950 1000 1050		
Mo 0–100	W	$6 \cdot 3 \times 10^{-4}$	80·5		1533–2260	IIa	119
Mo 0–10 0 to 10	Zr	1·6	107·3	$1 \cdot 3 \times 10^{-11}$ to $3 \cdot 7 \times 10^{-11}$	1650–1835 1835	IIa(ii) IIa(i)	115
(Mo₂Zr) 33⅓		$1 \cdot 10^{-3}$	55·6		820–1445	IIc	
N Sol. soln. range	Nb	0·061	38·8		800–1600	IIIa(i)	121
N Sol. soln. range	Th	0·0021	22·5		845–1490	IIIb(i)	122
N Conc. range at diff. temp. Sol. soln. range	Ti (α) (β)	0·012 0·035	45·25 33·8		900–1570 900–1570	IIIc(i) IIIa(i)	123
N Sol. soln. range	Zr (β)	0·015	30·7		920–1640	IIIa(i)	124
N Sol. soln. range	Zr (β) Hf $1 \cdot 8$–$2 \cdot 2$	0·003	33·6		900–1600	IIIa(i)	125
N Sol. soln. range	Zr (β) Sn (wt%) 1·8 2·6 5·0	0·011 0·014 0·011	31·4 30·9 29·4		1165–1640 1100–1530 1050–1490	IIIa(i)	126
Nb	Ti 0 20 40 60 80 100	*(a)* $2 \cdot 5 \times 10^{-3}$ $2 \cdot 5 \times 10^{-3}$ $3 \cdot 2 \times 10^{-3}$ $3 \cdot 8 \times 10^{-3}$ $3 \cdot 8 \times 10^{-3}$ $3 \cdot 8 \times 10^{-3}$	70 63 57 50 44 40	$D_{Ti} \sim 2 \times D_{Nb}$	1000–1590	IIa(i)	115

(a) Values taken from smoothed plots of A and Q against composition.

TABLE 4. CHEMICAL DIFFUSION COEFFICIENT MEASUREMENTS—*continued*

Element 1 at %	Element 2 at %	A	Q	D	Temp. range (°C)	Method	Ref.
Nb	U(γ)	(a)					
	2	$2 \cdot 8 \times 10^7$	148·9 ⎫		1500–1650 ⎫		
	12	$2 \cdot 3 \times 10^7$	144·2 ⎬				
	18	$9 \cdot 6 \times 10^6$	140·0		1400–1600		
	22	0·091	73·5 ⎫				
	28	0·113	72·9 ⎬		1300–1500		
	38	0·149	72·8		1150–1400		
	46	0·064	68·0		1150–1350		
	54	0·45	69·9		1075–1300 ⎬	IIa(i)	127
	62	0·84	68·5 ⎫				
	68	1·94	69·7 ⎬		950–1175		
	74	0·82	60·4 ⎫				
	78	1·16	60·3 ⎬		892–1125		
	82	$1 \cdot 19 \times 10^{-4}$	33·4		693–1025		
	93	$1 \cdot 63 \times 10^{-4}$	30·2 ⎫				
	97	$2 \cdot 31 \times 10^{-4}$	29·9 ⎬		693–1025 ⎭		
	97	$\left\{ \begin{array}{l} D_U - 3 \cdot 82 \times 10^{-3} \\ D_{Nb} - 7 \cdot 1 \times 10^{-3} \end{array} \right.$	29·8 ⎫ 39·3 ⎬				
	4 10–100			$D_{Nb} \sim 30 \times D_U$ $D_U > D_{Nb}$			

(a) This is a representative selection from a larger table of values in Ref. 127.

Nb	V	(b)					
	0	$1 \cdot 6 \times 10^{-2}$	98 ⎫				
	20	$1 \cdot 95 \times 10^{-2}$	82 ⎬				
	40	$2 \cdot 3 \times 10^{-2}$	70	$D_V \sim 3$ to $5 \times D_{Nb}$	1405–1750	IIa(i)	150
	60	$2 \cdot 8 \times 10^{-2}$	64				
	80	$3 \cdot 3 \times 10^{-2}$	63 ⎬				
	100	$3 \cdot 8 \times 10^{-2}$	63 ⎭				
Nb	W 10–90			*Figure 26* *Figure 27*	1900 ⎫ 2100 ⎬	IIa(i)	128
Nb 0 to 100	Zr	$\sim 10^{-2}$ to ~ 10	~ 48 to ~ 90		1445–1690	IIa(i)	115
Ni 0–3	Pb	$\sim 0 \cdot 66$	25·3		285–320	II	46
Ni 0–14·9	Pt	$7 \cdot 9 \times 10^{-4}$	43·1		1043–1401	IIa(ii)	28
Ni	Si 0–<1	1·5	61·7		1120–1300	IIa(ii)	112
Ni	Ti 0–0·9	0·86	61·4		1100–1300	IIa(ii)	23

(b) Values taken from smoothed plots of $\bar{A}$ and $\bar{Q}$ against composition.

Ni 0–sol. limit	U (γ)	2,500	46·0		850–1000	IIc	54
Ni	V 0–16·5	0·287	59·2		1100–1300	IIa(ii)	47
Ni	W 0–1·5 0–5	11·1 0·86	76·8 70·4		1150–1290 1100–1300	IIa(ii) IIa(ii)	23 47
O 0–1·13	Ta	0·015	26·7		700–1400	IIa(ii)	129
O Sol. soln. range Sol. soln. range	Ti (α) (β)	$5 \cdot 08 \times 10^{-3}$ 1·6	33·5 48·2	May vary significantly with conc.	700–850 950–1414	IIIa(i) IIIa(i)	130 123
O Sol. soln. range Sol. soln. range Sol. soln. range	Zr (α) (α) (Zircalloy) (β) (Zircalloy)	9·4 Varies by x 2 in different directions. quoted here 0·196 0·0453	51·78 (c) Average values of $\bar{A}$ and $\bar{Q}$ 41·0 (c) 28·2	Anisotropic	400–600 1000–1500 1000–1500	IIc IIIc(i) IIIa(i)	131 132 132

(c) The single equation $\bar{D}_\alpha = 5 \cdot 2 \exp(-50,800/RT)$ represents very well the combined results of Refs. 131 and 132 over the Temp. range 400°–1500°. Ref. 133 contains measurements which confirm this in the range of temperatures not covered by Refs. 131 and 132.

TABLE 4. CHEMICAL DIFFUSION COEFFICIENT MEASUREMENTS—*continued*

Element 1 at %	Element 2 at %	A	Q	D	Temp. range (°C)	Method	Ref.
Pb	Sn 0–2 Sol. soln. range	4·0 —	23·8 24·0	$\bar{D}$ increases with conc. of Sn.	245–300 170 and 181	IIa(ii) IIa(i)	10 135
Pb	Tl 0–2 0–53	0·025 1·03	19·4 24·6	Almost independent of conc.	220–285 260–315	IIa(ii) IIa(i)	39 113
Pt 2 50 55 65 77 80 85	W (β) $(\gamma)(a)$ $(\epsilon)(a)$ (α)	(b) $3\cdot1 \times 10^{2}$ $4\cdot7 \times 10^{-3}$ $3\cdot3 \times 10^{-3}$ $4\cdot4 \times 10^{-2}$ $1\cdot8 \times 10^{-2}$ $1\cdot2 \times 10^{-2}$ $1\cdot3 \times 10^{-2}$	139·0 83·6 82·1 92·0 78·0 75·4 74·2		1300–1743 1473–1743 1300–1743 1300–1700	IIa(i)	140
Pu 1·75 3·50 5·25 7·0 8·75 10·50 12·25 14·0 15·75	U (α)	$0\cdot14 \times 10^{-7}$ $0\cdot15 \times 10^{-7}$ $0\cdot18 \times 10^{-7}$ $0\cdot28 \times 10^{-7}$ $0\cdot44 \times 10^{-7}$ $0\cdot88 \times 10^{-7}$ $1\cdot18 \times 10^{-7}$ $2\cdot0 \times 10^{-7}$ $2\cdot57 \times 10^{-7}$	13·4 13·7 14·1 15·2 16·3 17·9 18·8 20·0 20·6	Probably a significant contribution to $\bar{D}$ from g.b. diffusion	410–540	IIa(i)	134

(a) The γ and ϵ are two new phases observed during the diffusion experiments and not previously reported.
(b) Arrhenius plots not always linear. $\bar{Q}$ and $\bar{A}$ derived from measurements at highest temperatures.

Element 1 at %	Element 2 at %	A	Q	D	Temp. range (°C)	Method	Ref.
Rh 3 60 70 90	W (α) (ϵ) (β)	$1\cdot3 \times 10^{-6}$ $1\cdot5 \times 10^{-6}$ $3\cdot1 \times 10^{-6}$ $2\cdot5 \times 10^{-6}$	58·0 41·7 43·4 41·6		1300–1800	IIa(i)	140
Ru 5 39 70 90	W (α) (σ) (β)	(c) $5\cdot5 \times 10^{-3}$ $1\cdot2 \times 10^{-5}$ $1\cdot8 \times 10^{-5}$ $1\cdot0 \times 10^{-5}$	93·5 61·0 49·5 57·2		1300–2025 1785–2025 1300–2025	IIa(i)	140
S ~0·01	Ni	$2\cdot3 \times 10^{6}$	90·0		1000–1200	IIIa(ii)	142
Si Sol. soln. range	U (γ)	20	45·0		850–1050	IIc	54
Sn 1·0 8·0 2·0	Ti (β) (β)	$8\cdot4 \times 10^{-7}$ $2\cdot7 \times 10^{-4}$	15·3 29·8	Increases linearly with C $\begin{cases} D_{Sn} = 9\cdot18 \times 10^{-9} \\ D_{Ti} = 2\cdot65 \times 10^{-9} \end{cases}$	1000–1250 1090–1250 1250	IIa(i)	26
Sn Sol. soln. range 0–3·9	Zr (α) (β)	$3\cdot10^{-6}$ $6\cdot9 \times 10^{-4}$	22·0 36·0		600–850 1100–1300	IIa(ii) IIa(ii)	136

(c) Arrhenius plots not always linear. $\bar{Q}$ and $\bar{A}$ derived from measurements at highest temperatures.

Element 1 at %	Element 2 at %	A	Q	D	Temp. range (°C)	Method	Ref.
Sr Sol. soln. range	U (γ)	$2\cdot38 \times 10^{-3}$	47·0		800–1000	IIc	109
Ti 10·0 20·0 30·0 40·0 50·0 60·0 70·0 80·0 90·0 95·0	U (γ)	$11\cdot10^{-3}$ $1\cdot4 \times 10^{-3}$ $1\cdot6 \times 10^{-3}$ $4\cdot0 \times 10^{-3}$ $9\cdot5 \times 10^{-3}$ $2\cdot6 \times 10^{-3}$ $2\cdot6 \times 10^{-3}$ $2\cdot2 \times 10^{-3}$ $1\cdot1 \times 10^{-3}$ $0\cdot46 \times 10^{-3}$	36·6 33·0 34·8 38·4 42·0 39·4 39·4 37·5 33·8 30·2		950–1075	IIa(i)	137

		D_{Ti}	D_U	
16·5		$5\cdot8 \times 10^{-9}$	$2\cdot2 \times 10^{-8}$	1075
18·0		$1\cdot2 \times 10^{-9}$	$4\cdot7 \times 10^{-8}$	950
		$2\cdot9 \times 10^{-9}$	$9\cdot5 \times 10^{-9}$	1000
		$4\cdot1 \times 10^{-9}$	$1\cdot6 \times 10^{-8}$	1050
16·5–18	$Q_U = 38\cdot5$; $Q_{Ti} = 40\cdot0$			

M (II)

TABLE 4. CHEMICAL DIFFUSION COEFFICIENT MEASUREMENTS—*continued*

Element 1 at %	Element 2 at %	A	Q	D	Temp. range (°C)	Method	Ref.
Ti (α)	V Sol. soln. range			$3 \cdot 91 \times 10^{-15}$ $4 \cdot 7 \times 10^{-15}$	600 } 700 }	IIa(ii) }	114
(β)	0–10	$1 \cdot 25 \times 10^{-2}$	$41 \cdot 4$		900–1300	IIa(i)	
(β)	2·0	$6 \cdot 0 \times 10^{-3}$	$39 \cdot 6$	Dep. on c in range 2–12% v. slight	900–1250	(IIa(i))	26
(β)	3·5			$\left\{ \begin{array}{l} D_{Ti} = 1 \cdot 31 \times 10^{-9} \\ D_V = 14 \cdot 9 \times 10^{-9} \end{array} \right\}$	1250		
Ti (α)	Zr 0–10	$1 \cdot 7 \times 10^{-12}$	$11 \cdot 8$		600–800	IIa(ii)	114
(β)	0–10	$1 \cdot 8 \times 10^{-2}$	$40 \cdot 1$		900–1300	IIa(ii)	114
U (β)	Xe 0–10⁻⁶	9×10^{-7}	23		700–750 }	IIIb(ii)	149
(γ)	0–10⁻⁶	10^8	98		810–1060 }		
U (γ)	Zr 10	$9 \cdot 5 \times 10^{-4}$	$32 \cdot 0$				
	20	$1 \cdot 3 \times 10^{-4}$	$28 \cdot 6$				
	30	$0 \cdot 35 \times 10^{-4}$	$26 \cdot 3$				
	40	$0 \cdot 4 \times 10^{-4}$	$27 \cdot 4$				
	50	$0 \cdot 8 \times 10^{-4}$	$29 \cdot 7$		950–1075	IIa(i)	138
	60	$0 \cdot 63 \times 10^{-4}$	$29 \cdot 7$				
	70	$0 \cdot 55 \times 10^{-4}$	$29 \cdot 7$				
	80	$3 \cdot 2 \times 10^{-4}$	$34 \cdot 3$				
	90	78×10^{-4}	$41 \cdot 0$				
	95	870×10^{-4}	$47 \cdot 0$				
	10–95			D_U and D_{Zr} Figures 28 and 29	950–1040	IIa(i)	139

REFERENCES TO TABLE 4

1. A. D. Le Claire and A. H. Rowe, *Rev. Métall.*, 1955, **52**, 94.
2. Th. Heumann and S. Dittrich, *Z. Electrochem*, 1957, **61**, 1138.
3. H. Ebert and G. Trommsdorf, *ibid.*, 1950, **54**, 294.
4. R. W. Balluffi and L. L. Seigle, *J. appl. Phys.*, 1954, **25**, 607.
5. W. A. Johnson, *Trans. A.I.M.E.*, 1942, **147**, 331.
6. O. Kubaschewski, *Trans. Faraday Soc.*, 1950, **46**, 713.
7. W. Eichenauer, H. Kunzi and A. Pebler, *Z. Metallk.*, 1958, **49**, 220.
8. J. M. Tobin, *Acta. metall.*, 1957, **5**, 398.
9. W. Eichenauer and G. Müller, *Z. Metallk.*, 1962, **53**, 321; 1962, **53**, 700.
10. W. Seith and J. G. Laird, *ibid.*, 1932, **24**, 193.
11. T. Heumann and P. Lohmann, *Z. Electrochem.*, 1955, **59**, 849.
12. A. G. Guy, *Trans. Metall. Soc., A.I.M.E.*, 1959, **215**, 279.
13. J. M. Tobin, *Acta metall.*, 1959, **7**, 701.
14. H. Buckle and J. Descamps, *Rev. Métall.*, 1951, **48**, 569.
15. J. B. Murphy, *Acta metall.*, 1961, **9**, 563.
16. L. C. Correa da Silva and R. F. Mehl, *Trans. Am. Inst. min. Engrs*, 1951, **191**, 155.
17. M. K. Asundi and D. R. F. West, *J. Inst. Metals*, 1964, **92**, 428.
18. K. Sato, *Trans. Jap. Inst. Met.*, 1963, **4**, 121.
19. W. Eichenauer and A. Pebler, *Z. Metallk.*, 1957, **48**, 373.
20. L. P. Costas, *U.S.A. Rep.*, TID-16676, 1962.
21. H. Bückle, *Z. Electrochem.*, 1943, **49**, 238.
22. C. E. Ransley and H. Neufeld, *J. Inst. Metals*, 1950, **78**, 25.
23. R. A. Swalin and A. Martin, *J. metal. Trans. A.I.M.E.*, 1956, **206**, 567.
24. A. Beerwald, *Z. Electrochem.*, 1939, **45**, 789.
25. R. F. Mehl, F. N. Rhines and K. A. von den Steiner, *Metals Alloys*, 1941, **13**, 41.
26. D. Goold, *J. Inst. Metals*, 1960, **88**, 444.
27. J. E. Hilliard, B. L. Averbach and M. Cohen, *Acta metall.*, 1959, **7**, 86.
28. O. Kubeschewski and H. Ebert, *Z. Electrochem.*, 1944, **50**, 138.
29. W. Eichenauer and D. Liebscher, *Z. Naturforsch.*, 1962, **17a**, 355.
30. G. W. Powell and J. D. Braun, *Trans. metall. Soc. A.I.M.E.*, 1964, **230**, 694.
31. J. E. Reynolds, B. L. Averbach and M. Cohen, *Acta metall.*, 1957, **5**, 29.
32. W. Seith and K. Etzold, *Z. Electrochem.*, 1934, **40**, 829; 1935, **41**, 122.
33. W. Jost, *Z. phys. Chem.*, 1933, **B21**, 158.
34. A. Bolk, *Acta metall.*, 1961, **9**, 643.
35. P. E. Busby and C. Wells, *J. Metals*, 1954, **6**, 972.
36. ——, M. E. Warga and C. Wells, *ibid.*, 1953, **5**, 1463.
37. R. Reinbach and F. Krietsh, *Z. Metallk.*, 1963, **54**, 173.
38. R. Le Hazif, G. Donze, J. M. Dupouy and Y. Adda, *Mem. Sci. Rev. Met.*, 1964, **LXI**, 467.
39. W. Seith and F. G. Laird, *Z. Metallk.*, 1932, **24**, 193.
40. C. G. Homan, *Acta metall.*, 1964, **12**, 1071.
41. R. P. Smith, *Trans. metall. Soc. A.I.M.E.*, 1964, **230**, 476.
42. C. Wells, W. Batz and R. F. Mehl, *Trans. Am. Inst. min. Engrs*, 1950, **188**, 553.
43. J. J. Lander, H. E. Kern and A. L. Beach, *J. appl. Phys.*, 1952, **23**, 1305.
44. D. T. Peterson, *Trans. Am. Soc. Metals*, 1961, **53**, 765.
45. F. C. Wagner, E. J. Burcur and M. A. Steinberg, *ibid.*, 1956, **48**, 742.
46. W. Seith, E. Hofer and H. Etzold, *Z. Electrochem.*, 1934, **40**, 322.
47. A. Davin, V. Leroy, D. Coutsouradis and L. Habraken, *Rev. Métall.*, 1963, **60**, 275; *Cobalt*, June 1963, **19**.
48. J. W. Weeton, *Trans. Am. Soc. Metals*, 1952, **44**, 436.
49. T. Heumann and H. Bohmer, *Arch. Eisenhütt Wes.*, 1960, **31**, 749.
50. H. W. Paxton and E. J. Pasierb, *Trans. metall. Soc. A.I.M.E.*, 1960, **218**, 794.

51. J. R. Manning, *Phys. Rev.*, 1959, **116**, 69.
52. A. E. Austin and N. A. Richard, *J. appl. Phys.*, 1962, **33**, 3569.
53. R. F. Peart and D. H. Tomlin, *J. Phys. Chem. Solids*, 1962, **23**, 1169.
54. M. Mossé, V. Levy and Y. Adda, *C.R. Acad. Sci. Paris*, 1960, **250**, 3171.
55. D. E. Thomas and C. E. Birchenall, *Trans. Am. Inst. min. Engrs*, 1952, **194**, 867.
56. A. E. Austin and N. A. Richard, *J. appl. Phys.*, 1961, **32**, 1462.
57. C. Matano, *Jap. J. Phys.*, 1933, **8**, 109.
58. C. E. Ransley, *J. Inst. Metals*, 1939, **65**, 147.
59. F. N. Rhines and R. F. Mehl, *Trans. Am. Inst. min. Engrs*, 1938, **128**, 185.
60. E. Starke and H. Wever, *Z. Metallk.*, 1964, **55**, 107.
61. G. T. Horne and R. F. Mehl, *Trans. Am. Inst. min. Engrs*, 1955, **203**, 88.
62. R. Resnick and R. W. Balluffi, *ibid.*, 1955, **203**, 1004.
63. U. S. Landergren, C. E. Birchenall and R. F. Mehl, *ibid.*, 1956, **206**, 73.
64. R. F. Mehl and C. F. Lutz, *Trans. metall. Soc. A.I.M.E.*, 1961, **221**, 561.
65. E. W. Johnson and M. L. Hill, *ibid.*, 1960, **218**, 1104.
66. W. Geller and T. H. Sun, *Arch. Eisenhütt Wes.*, 1950, **21**, 423.
67. C. Wells and R. F. Mehl, *Trans. Am. Inst. min. Engrs*, 1941, **145**, 315.
68. P. Grieveson and E. T. Turkdogan, *Trans. metall. Soc. A.I.M.E.*, 1964, **230**, 1604.
69. J. D. Fast and M. B. Verrijp, *J. Iron St. Inst.*, 1954, **176**, 24.
70. P. Grieveson and E. T. Turkdogan, *Trans. metall. Soc. A.I.M.E.*, 1964, **230**, 411.
71. L. S. Darken, R. P. Smith and E. W. Filer, *Trans. Am. Inst. min. Engrs*, 1951, **191**, 1174.
72. C. Wells and R. F. Mehl, *ibid.*, 1941, **145**, 129.
73. N. G. Ainslie and A. E. Seybolt, *J. Iron St. Inst.*, 1960, **194**, 341.
74. G. Seibel, *C.R. Acad. Sci. Paris*, 1962, **255**, 3182; *Mem. Sci. Rev. Metall.*, 1964, **61**, 413.
75. W. Batz, H. W. Mead and C. E. Birchenall, *Trans. metall. Soc. A.M.I.E.*, 1952, **194**, 1070.
76. F. J. Bradshaw, G. Hoyle and K. Speight, *Nature, Lond.*, 1953, **171**, 488.
77. S. H. Moll and R. E. Ogilvie, *Trans. metall. Soc. A.I.M.E.*, 1959, **215**, 613.
78. R. E. Hannemann, R. E. Ogilvie and H. C. Gates, *Trans. metall. Soc. A.I.M.E.*, 1965, **233**, 691.
79. G. Seibel, *C.R. Acad. Sci. Paris*, 1963, **256**, 4661; *Mem. Sci. Rév. Met.*, 1964, **61**, 413.
80. E. Fitzer, *Z. Metallk.*, 1953, **44**, 462.
81. C. O. Bannister and W. D. J. Jones, *J. Iron St. Inst.*, 1931, **124**, 71.
82. J. A. M. van Liempt, *Rec. Trav. Chim. Pays Bas*, 1945, **64**, 239.
83. G. Grube and K. Schneider, *Z. anorg. Chem.*, 1927, **168**, 17.
84. R. C. Frank and J. E. Thomas, *J. Phys. Chem. Solids*, 1960, **16**, 144.
85. J. Tournier, *Rep. CEA-R-2446*, October 1964.
86. J. P. Pemsler and E. J. Rapperport, *Trans. metall. Soc. A.I.M.E.*, 1964, **230**, 90.
87. W. M. Albrecht, W. D. Goode and M. W. Mallet, *J. Electrochem. Soc.*, 1959, **106**, 981.
88. M. L. Hill and E. W. Johnson, *Acta metall.*, 1955, **3**, 566.
89. J. L. Ham, *Trans. Am. Soc. Metals*, 1945, **35**, 331.
90. A. G. Edwards, *Brit. J. appl. Phys.*, 1957, **8**, 406.
91. M. van Sway and C. E. Birchenall, *Trans. metall. Soc. A.I.M.E.*, 1960, **218**, 285.
92. W. D. Davis, *U.S. Rep.* K.A.P.L. 1227, October, 1954.
93. O. M. Katz and E. A. Gulbransen, *Rev. Sci. Inst.*, 1960, **31**, 615.
94. W. D. Davis, *U.S. Rep.* K.A.P.L. 1375, April 1955.
95. O. N. Salmon, D. Randall and E. A. Wilk, *ibid.*, 1674, November 1956; K.A.P.L. 984, May 1954.
96. D. T. Peterson and D. G. Westlake, *J. phys. Chem.*, 1960, **64**, 649.
97. R. J. Wasilewski and G. L. Kehl, *Metallurgica*, 1954, **50**, 225.
98. M. W. Mallet and M. J. Trzeciak, *Trans. Am. Soc. Metals*, 1958, **50**, 981.
99. H. W. Meyers, J. W. Varwig, J. L. Marshall, L. G. Weber and J. E. Kenelley, *U.S.A.E.C. Rep.* MCW-1439 December 1959.
100. M. Someno, *Nippon Kink. Gakk.*, 1960, **24**, 249.
101. M. W. Mallet and M. W. Albrecht, *J. Electrochem. Soc.*, 1957, **104**, 142.
102. A. Sawatzky, *J. nucl. Mater.*, 1960, **2**, 62.
103. V. L. Gelezunas, *Thesis*, University of Cincinnati, Ohio, 1962.
104. H. R. Glyde, *Phil. Mag.*, 1965, **12**, 919.
105. ——, *ibid.* 1965, **12**, 997.
106. A. van Wieringen and N. Warmoltz, *Physica*, 1956, **22**, 849.
107. A. Andrew, C. R. Davidson and L. E. Glasgow, *U.S. Rep.* NAA-SR-1598, 1956.
108. J. P. Pemsler, *J. Electrochem. Soc.*, 1964, **111**, 1185.
109. Y. Adda, V. Levy, Z. Hadari and J. Tournier, *Rev. Métall.*, 1959, **57**, 278.
110. E. M. Pell, *Phys. Rev.*, 1960, **119**, 1014.
111. H. M. Love and G. M. McCracken, *Can. J. Phys.*, 1963, **41**, 83.
112. R. A. Swalin, A. Martin and R. Olsen, *Trans. Am. Inst. min. Engrs*, 1957, **209**, 936.
113. W. Seith and J. Herrmann, *Z. Electrochem*, 1940, **46**, 213.
114. R. P. Elliot, *U.S. Rep.* AD.290336, March 1962.
115. C. S. Hartley, J. E. Steedly and L. D. Parsons, *U.S. Rep.* ML-TDR-64-316, December 1964.
116. Y. Adda and J. Philibert, *C.R. Acad. Sci. Paris*, 1958, **246**, 113; Rep. C.E.A.-880, March 1958.
117. A. M. Rodin and V. V. Surenyants, *Phys. Metals Metallogr.*, 1960, **10** (2), 58.
118. D. Calais, M. Beyeler, M. Mouchnino, A. van Craeynest and Y. Adda, *C.R. Aacd. Sci. Paris*, 1963, **257**, 1285.
119. J. A. M. van Liempt, *Rec. Trav. Chim. Pays Bas*, 1932, **51**, 114.
120. P. Gröbner, *Hutnické listy*, 1955, **10**, 200.
121. W. M. Albrecht and W. D. Goode, *U.S. Rep.* BMI-1360, 1959.
122. A. F. Gerds and M. W. Mallett, *J. Electrochem. Soc.*, 1954, **101**, 175.
123. R. J. Wasilewski and G. L. Kehl, *J. Inst. Metals*, 1954, **88**, 94.
124. M. W. Mallett, J. Belle and B. B. Cleland, *J. Electrochem. Soc.*, 1954, **101**, 1.
125. ——, E. M. Baroody, H. R. Nelson and C. A. Papp, *J. Electrochem. Soc.*, 1953, **100**, 103.
126. ——, J. Belle and B. B. Cleland, *U.S. Rep.* BMI-829, May 1953.
127. N. L. Peterson and R. E. Ogilvie, *Trans. metall. Soc. A.I.M.E.*, 1963, **227**, 1083.
128. E. Gebhardt and K. Kirner, *Z. Metallk*, 1963, **54**, 437; *Z. Erzb. Metallhutt.*, 1963, **16**, 698.
129. ——, H. D. Seghezzi and A. Stegherr, *ibid.*, 1957, **48**, 624.
130. W. P. Roe, H. R. Palmer and W. R. Opie, *Trans. Am. Soc. Metals*, 1960, **52**, 191.
131. J. P. Pemsler, *J. Electrochem. Soc.*, 1958, **105**, 315.
132. M. W. Mallett, M. W. Albrecht and P. R. Wilson, *ibid.*, 1959, **106**, 181.
133. G. Béranger, *C.R. Acad. Sci. Paris*, 1964, **259**, 4663.
134. M. Dupuy and D. Calais, *Mem. Sci. Rev. Met.*, 1965, **LXII**, 721.
135. H. Cordus and M. Kukuk, *Z. anorg. Allgem. Chemie*, 1960, **306**, 121.
136. R. Resnick and R. Balluffii, *U.S. Rep.* S.E.P. 118, August, 1953.
137. Y. Adda and J. Philibert, *Acta metall.*, 1960, **8**, 700.
138. ——, —— and Faraggi, *Rev. Métall.*, 1957, **54**, 597.
139. ——, C. Mairy and J. L. Andreu, *ibid.*, 1960, **57**, 550.
140. E. J. Rapperport, V. Merses and M. F. Smith, *U.S. Rep.* ML-TDR-64-61 (March 1964).
141. J. I. Goldstein, R. E. Hanneman and R. E. Ogilvie, *Trans. metall. Soc. A.I.M.E.*, 1965, **233**, 812.
142. I. Pfeiffer, *Z. Metallk.*, 1955, **46**, 516.
143. L. S. DeLuca, *U.S. Rep.* KAPL-M-LSD-1, August, 1960.
144. T. Heumann and A. Kottmann, *Z. Metallk.*, 1953, **44**, 139.
145. M. A. Krishtal, *Dokl. Akad. Nauk. S.S.S.R.*, 1953, **92**, 951, and Nsf-tr 223.

146. M. Blanter, *Zhur. Tech. Phys. S.S.S.R.*, 1950, **20**, 1001.
147. ——, *ibid.*, 1950, **20**, 217.
148. ——, *ibid.*, 1951, **21**, 818.
149. M. B. Perraillon, V. Levy and M. Y. Adda, *Commn.* to 1964 Autumn Meeting, Société Française de Métallurgie.
150. R. C. Reiss, C. S. Hartley and J. E. Steedly, *J. less-common Metals*, 1965, **9**, 309

TABLE 4(a). CHEMICAL DIFFUSION IN TERNARY SYSTEMS

System	Symbol	Comp. range (wt%)		Temp. (°C)	Ref.

Al
  Zn
  Cu
	1
	
	2
		11·8–12·6 Zn
		0–3·66 Cu
		0–11·8 Zn
		3·32–3·66 Cu
			$D_{11} = 2\cdot91 \times 10^{-9}$ $D_{12} = 1\cdot65 \times 10^{-9}$
			$D_{22} = 1\cdot23 \times 10^{-9}$ $D_{12} = \quad —$
			$D_{11} = 3\cdot4 \times 10^{-9}$ $D_{21} = \quad —$
			$D_{22} = (0\cdot454 + 0\cdot067(\%\mathrm{Zn}))10^{-9}$ $D_{21} = 0$
				} 504	1

Al
  Cu
  Mn
	1
	2
		3·7–3·8
		0–0·66
			$D_{22} = 9\cdot10^{-12}$ $D_{21} = \quad —$
			$D_{11} = 1\cdot78 \times 10^{-9}$ $D_{12} = -0\cdot8 \times 10^{-9}$
				} 557	2

Fe
  C
  Si
	1
	2
		0·441–0·478
		0·05–3·89
			$D_{11} = 4\cdot8 \times 10^{-7}$ $D_{12} = 0\cdot34 \times 10^{-7}$
			$D_{22} = 1\cdot0 \times 10^{-8}$ $D_{21} = 1\cdot4 \times 10^{-7}$
				} 1050	3 and 4

REFERENCES TO TABLE 4(a)

1. J. S. Kirkaldy, Zia-Ul-Haq and L. C. Brown, *Trans. Am. Soc. Metals*, 1963, **56**, 835.
2. ——, G. R. Mason and W. J. Slates, *Trans. Can. Inst. Min. Metall.*, 1961, **64**, 53.
3. L. S. Darken, *Trans. Am. Inst. Min. Engrs*, 1949, **180**, 430.
4. J. S. Kirkaldy, *Can. J. Phys.*, 1958, **36**, 899.

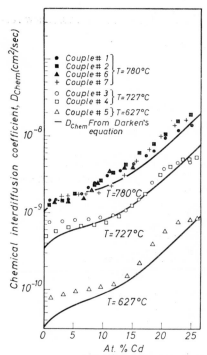

Figure 3. *Chemical diffusion coefficients in Ag Cd alloys* [51]

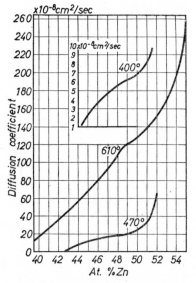

Figure 4. *Chemical diffusion coefficients in Ag Zn alloys* [11]

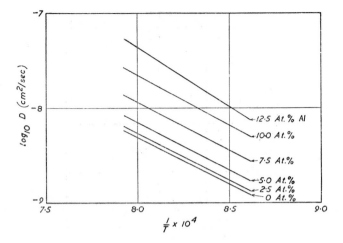

Figure 5. *Chemical diffusion coefficients in* Cu Al *alloys* [16]

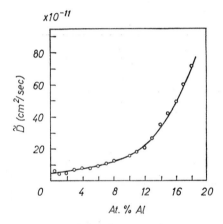

Figure 6. *Chemical diffusion in* Al Fe *alloys at* 850°C [18]

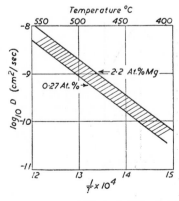

Figure 7. *Chemical diffusion in* Al Mg *alloys* [21]

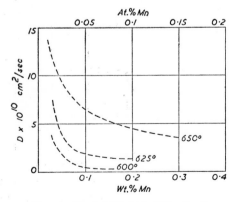

Figure 8. *Chemical diffusion in* Al Mn *alloys* [21]

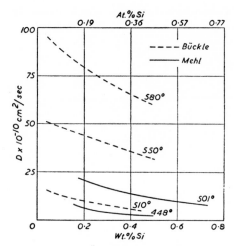

Figure 9.   Chemical diffusion in Al Si alloys [21], [25]

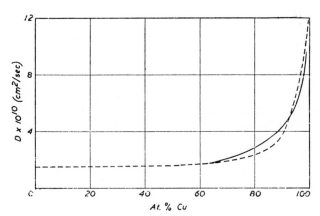

Figure 10.   Chemical diffusion in Ni Cu *system at* 1025°C [55], [57]

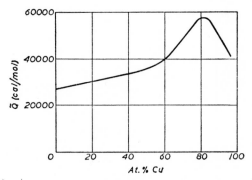

Figure 11.   Q̄ for diffusion in Ni Cu *system* [18]

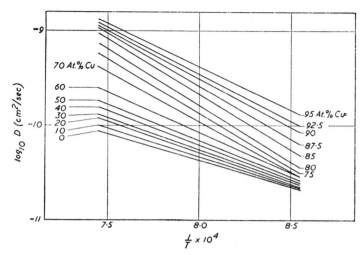

Figure 12. $\tilde{D}$ *for diffusion in* Ni Cu *system* [16]

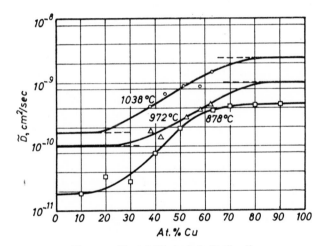

Figure 13. *Chemical diffusion in* Cu Pd *alloys* [55]

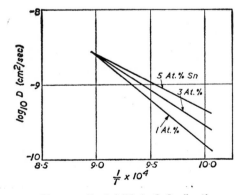

Figure 14. *Chemical diffusion* Cu Sn *alloys* [16]

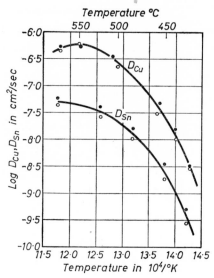

Figure 15. *Partial diffusion coefficients for* Cu *and* Sn *in* δ-*phase* Cu Sn *alloys* [60]

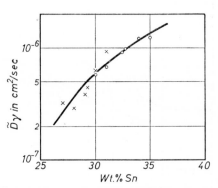

Figure 16. *Chemical diffusion coefficients for* γ-*phase of* Cu Sn *system at* 706·5°C [60]

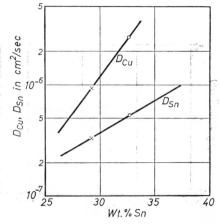

Figure 17. *Partial diffusion coefficients for* Cu *and* Sn *in* γ Cu Sn *alloys at* 706·5°C [60]

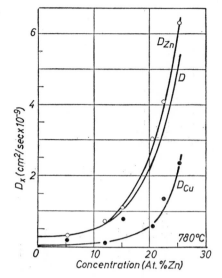

Figure 18. *Chemical and partial diffusion coefficients in α Cu Zn system at 780°C* [61]

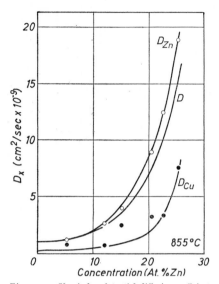

Figure 19. *Chemical and partial diffusion coefficients in α Cu Zn system at 855°C* [61]

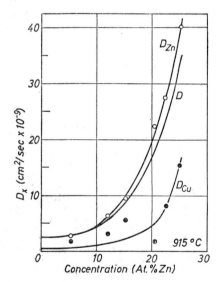

Figure 20. *Chemical and partial diffusion coefficients in α Cu Zn system at 915°C* [61]

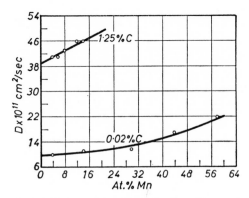

*Figure 21.   Chemical diffusion coefficients in γ Fe Mn alloys with 0·02 and 1·25 wt% C.
Temperature 1200°C* [67]

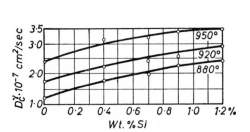

*Figure 22.   Effect of Si content on chemical diffusion of C in
γ Fe* [146]

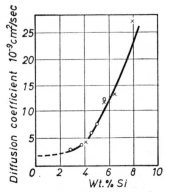

*Figure 23.   Chemical diffusion coefficients in
α Fe Si alloys at 1150°C* [80]

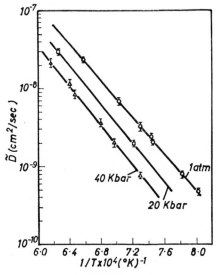

*Figure 24.   Chemical diffusion in α-Fe V 10% alloy at 1,
20 and 40 k bar pressure* [78]

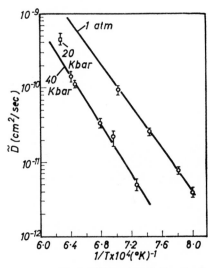

*Figure 25.   Chemical diffusion in γ Fe V 0·7% alloy
at 1, 20 and 40 k bar pressure* [78]

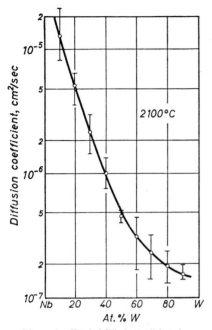

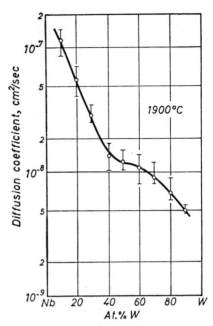

Figure 26.　*Chemical diffusion coefficients in*
Nb W *system at 1900°C* [128]

Figure 27.　*Chemical diffusion coefficients in*
Nb W *system at 2100°C* [128]

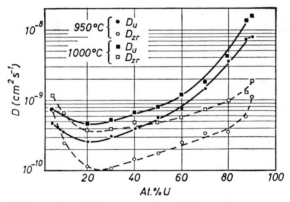

Figure 28.　*Partial diffusion coefficients* D_U *and* D_{Zr} *in* γ U Zr *alloys* [139]

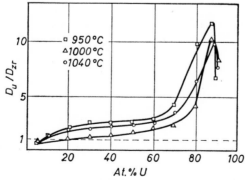

Figure 29.　*Ratio of partial diffusion coefficients in* γ U Zr *alloys* [139]

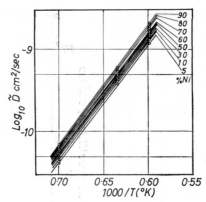

Figure 30. Chemical diffusion coefficients in Co–Ni alloys [144]

TABLE 5. GRAIN BOUNDARY SELF-DIFFUSION COEFFICIENTS

Element	A_{gb}	Q_{gb}	Temp. range (°C)	Method	Ref.
Ag	0·12	21·5	} 350–480 {	99·999% } Polycrystals	1
	0·03	20·2		99·97% }	
	(D_p) 0·14	19·7	400–525	99·98% Bicrystals-tilt boundaries 9–28°	2
	(D_p) 0·0032	18·0	420–520	99·999% Bicrystal-twist boundaries 2–30°	3
			D_{gb} is anisotropic within a boundary.		4
Zn	0·22	14·3 ± 0·2	} 75–160 {	99·999% } Polycrystals	5
	0·38	14·6 ± 0·2		99·99% }	
	—	12·3			6
Cd	1·0	13·0 ± 1·2	50–110	99·5% Polycrystals	7
Sn	$6·44 {+5·9 \atop -4·9} \times 10^{-2}$	9·55 ± 0·7	40–115	99·99% Polycrystals	8
Pb	0·81	15·7	214–260	Polycrystals	9
			D_{gb} is anisotropic within a boundary.		10
Cr	—	46	1000–1350	Polycrystals	11
Fe	(α) 13·0	40	580–660	99·7% } Polycrystals	12
	2·5	40	530–650	99·96% }	
	(γ) 5·0	39	920–1020	99·7% } Polycrystals	13
	3·4	39		99·99% }	
Co	0·15	39 ± 0·5			14
Ni	$1·75 {+2·1 \atop -1·0} \times 10^{-2}$	28·2 ± 2·0	850–1100	99·85% Polycrystals	15
	—	26·0 ± 1·5 (20° < θ < 70°)	700–1100	99·95% Bicrystals-tilt boundaries D_{gb}/D_v varies from ~5·10³ for θ = 5° boundaries at 1100°C to ~2·10⁷ for 45° boundaries at 700°C	} 16

All values of A_{gb} are calculated assuming a grain boundary width $\delta = 5 \cdot 10^{-8}$ cm. (D_p) signifies that A_{gb} refers to dislocation pipe diffusion. See Ref. 2.

REFERENCES TO TABLE 5

1. R. E. Hoffman and D. Turnbull, *J. appl. Phys.*, 1951, **22**, 634.
2. D. Turnbull and R. E. Hoffman, *Acta metall.*, 1954, **2**, 419.
3. G. Love and P. G. Shewmon, *ibid.*, 1963, **11**, 899.
4. R. E. Hoffman, *ibid.*, 1956, **4**, 97.
5. E. S. Wajda, *ibid.*, 1954, **2**, 184.
6. F. Voigtmann, *Thesis*, Dresden, 1961.
7. E. S. Wajda, G. A. Shirn and H. B. Huntington, *Acta metall.*, 1955, **3**, 39.
8. W. Lange and D. Bergner, *Phys. Stat. Sol.*, 1962, **2**, 1410.
9. B. Okkerse, *Acta metall.*, 1954, **2**, 551.
10. ——, T. J. Tiedema and W. G. Burgers, *ibid.*, 1955, **3**, 300.
11. S. Z. Bokstein, S. T. Kishkin and L. M. Moroz, *UNESCO Int. Conf. Rad. Isotopes and Sci. Res.*, 1957, Pap. 193.
12. C. Leymonie and P. Lacombe, *Mém. scient. Revue Métall.*, 1960, **57**, 285.
13. P. Guiraldenq and P. Lacombe, *Acta metall.*, 1965, **13**, 51.
14. S. D. Gerzricken, T. K. Yatsenko and L. Slastnikov, *Vaprosy fiziki metallov i metallovedeniva*, Kiev, 1959, **9**, 154.
15. W. Lange, A. Hässner and G. Mischer, *Phys. Stat. Sol.*, 1964, **5**, 63.
16. W. R. Upthegrove and M. J. Sinnott, *Trans. Am. Soc. Metals*, 1958, **50**, 1031.

TABLE 6.   SELF-DIFFUSION COEFFICIENTS IN LIQUID METALS

Element	A	Q	Temp. range (°C)	Method	Ref.
Na²²	$(1\cdot10 \pm 0\cdot37) \times 10^{-3}$	$2\cdot43 \pm 0\cdot2$	98–226	Capilliary	1
K⁴²	$(1\cdot67 \pm 0\cdot55) \times 10^{-3}$	$2\cdot55 \pm 0\cdot27$	67–217	Capilliary	7
Cu⁶⁴	$(1\cdot46 \pm 0\cdot01) \times 10^{-3}$	$9\cdot71 \pm 0\cdot71$		Capilliary	2
Ag¹¹⁰	$(7\cdot10 \pm 0\cdot06) \times 10^{-4}$ $(5\cdot8 \pm 1\cdot4) \times 10^{-4}$	$8\cdot15 \pm 1\cdot13$ $7\cdot66 \pm 0\cdot67$	1000–1105 957–1350	Capilliary Capilliary	3 4
Zn⁶⁵	$8\cdot2 \times 10^{-4}$ 12×10^{-4}	$5\cdot09$ $5\cdot60$	450–600 420–600	Capilliary Capilliary	5 6
Hg	$0\cdot85 \times 10^{-4}$ $(3\cdot01 \pm 0\cdot1) \times 10^{-4}$ $1\cdot10 \times 10^{-4}$	$1\cdot005 \pm 0\cdot092$ $1\cdot80 \pm 0\cdot42$ $1\cdot15$	0–98 30–80 23–60	IIa(ii) (Shear cell) Capilliary IIa(ii)	8 9 10
Ga	$1\cdot07 \times 10^{-4}$	$1\cdot122$	30–100	IIa(ii) (Shear cell)	11
In	$(2\cdot89 \pm 0\cdot25) \times 10^{-4}$ $(3\cdot34 \pm 0\cdot21) \times 10^{-4}$	$2\cdot43 \pm 0\cdot50$ $2\cdot554 \pm 0\cdot079$	170–750	Capilliary IIa(ii)	12 13
Tl	$9\cdot45 \times 10^{-4}$	$4\cdot4$	(Estimated from viscosity)		14
Sn	$(3\cdot24 \pm 0\cdot12) \times 10^{-4}$ $(2\cdot04 \pm 0\cdot5) \times 10^{-4}$	$2\cdot768 \pm 0\cdot08$ $2\cdot00 \pm 0\cdot20$	267–683 250–475	IIa(ii) IIb	13 15
Pb	$(9\cdot15 \pm 0\cdot3) \times 10^{-4}$	$4\cdot45 \pm 0\cdot33$	606–930	Capilliary	16

REFERENCES TO TABLE 6

1. R. E. Meyer and N. H. Nachtrieb, *J. chem. Phys.*, 1955, **23**, 1851.
2. J. Henderson and L. Yang, *Trans. metall. Soc. A.I.M.E.*, 1961, **221**, 72.
3. L. Yang, S. Kado and G. Derge, *ibid.*, 1958, **212**, 628.
4. V. G. Leak and R. A. Swalin, *ibid.*, 1964, **230**, 426.
5. N. H. Nachtrieb, E. Fraga and C. Wahl, *J. phys. Chem.*, 1963, **67**, 2353.
6. W. Lange, W. Pippel and F. Bendel, *Z. Phys. Chem.*, 1959, **212**, 238.
7. J. Röhlin and A. Lodding, *Z. Naturf.*, 1962, **17**, 1081.
8. N. H. Nachtrieb and J. Petit, *J. chem. Phys.*, 1956, **24**, 746.
9. T. F. Kassner, R. J. Russell and R. E. Grace, *Trans. Am. Soc. Metals*, 1962, **55**, 858.
10. D. S. Brown and D. G. Tuck, *Trans. Faraday Soc.*, 1964, **60**, 1230.
11. J. Petit and N. H. Nachtrieb, *J. chem. Phys.*, 1956, **24**, 1027.
12. A. Lodding, *Z. Naturf.*, 1956, **11a**, 200.
13. G. Careri, A. Paoletti and M. Vincentini, *Il Nuovo Cim.*, 1958, **10**, 1088.
14. J. A. Cahill and A. V. Grosse, *J. phys. Chem.*, 1965, **69**, 518.
15. W. Lange, W. Pippel and H. Opperman, *Isotopen-Tech.*, 1962, **2**, 132.
16. S. J. Rothman and L. D. Hall, *Trans. Am. Inst. Min. Engrs.*, 1956, **206**, 199.

INDEX

[1]